Architecture Competition

Much valued by design professionals, controversially discussed in the media, regularly misunderstood by the public and systematically regulated by public procurement; in recent years, architecture competitions have become projection screens for various and often incommensurable desires and hopes.

Almost all texts on architectural competition engage it for particular reasons, whether these be for celebration of the procedure, or dismissal. Moving on from such polarised views, *Architecture Competition* is a revelatory study on what really happens when competitions take place. But the story is not just about architecture and design; it is about the whole construction process, from the definition of the spatial programme, to judgement and selection of projects and the realization of the building.

This book explores the competition in the building process as it takes place, but also before and after its execution. It demonstrates that competitions are not just one step of many to be taken, but that competitive design procedures shape the entire process. Along the way the book exposes, among others, one of the key evolutions of design competitions – that competition procedures need to be regulated in order to respond to public awarding rules and need to integrate an increasing amount of given standards regarding, for example, energy efficiency, fire safety and thermal comfort. These notions force competing architects to respond to inflexible and overloaded competition programmes instead of focusing on genuinely crafting an architectural project.

If the architecture competition wants to be more highly valued as a design tool, it should pay attention to the iterative nature of design and to the fact that perspectives on the problem to be solved often change in process.

Ignaz Strebel and **Jan Silberberger** are senior researchers within the Center for Research on Architecture, Society and the Built Environment, Department of Architecture, ETH Zurich. They have collaborated on various research and design projects, including the development of the internet platform KONKURADO | Web of Design Competitions 1.0.

Design and the Built Environment
Series editor: Matthew Carmona

This series provides a means to disseminate substantive research in urban design and its allied fields. Contributions are welcomed which are the result of original empirical research, scholarly evaluation, reflection on the practice and the process of urban design, critical analysis of particular aspects of the built environment, or important conference proceedings. Volumes should be of international interest, although they may focus on the particular experience and practice in one country. They may reflect theory and practice from across one or more of the spatial scales over which urban design operates.

Total Landscape, Theme Parks, Public Space
Miodrag Mitrasinovic

The Design of Frontier Spaces
Control and ambiguity
Edited by Carolyn Loeb and Andreas Luescher

Urban Design in the Arab World
Reconceptualizing boundaries
Edited by Robert Saliba

Towns and Cities: Function in Form
Urban structures, economics and society
Julian Hart

Urban Planning in Lusophone African Countries
Edited by Carlos Nunes Silva

Urban Planning in North Africa
Edited by Carlos Nunes Silva

Spatial Cultures
Towards a new social morphology of cities past and present
Edited by Sam Griffiths and Alexander von Lünen

The Social Fabric of Cities
Vinicius Netto

The Ethics of a Potential Urbanism
Critical encounters between Giorgio Agamben and architecture
Camillo Boano

Architecture Competition
Project Design and the Building Process

Ignaz Strebel and Jan Silberberger

With contributions from
Camille Crossman, Peter Holm Jacobsen,
Andreas Kamstrup, Kristian Kreiner,
Torsten Schmiedeknecht and Leentje Volker

Routledge
Taylor & Francis Group

LONDON AND NEW YORK

First published 2017
by Routledge
2 Park Square, Milton Park, Abingdon, Oxon OX14 4RN

and by Routledge
711 Third Avenue, New York, NY 10017

Routledge is an imprint of the Taylor & Francis Group, an informa business

© 2017 selection and editorial matter, Ignaz Strebel and Jan Silberberger, individual chapters, the contributors

The right of Ignaz Strebel and Jan Silberberger to be identified as the authors of the editorial material, and of the authors for their individual chapters, has been asserted in accordance with sections 77 and 78 of the Copyright, Designs and Patents Act 1988.

All rights reserved. No part of this book may be reprinted or reproduced or utilised in any form or by any electronic, mechanical, or other means, now known or hereafter invented, including photocopying and recording, or in any information storage or retrieval system, without permission in writing from the publishers.

Trademark notice: Product or corporate names may be trademarks or registered trademarks, and are used only for identification and explanation without intent to infringe.

British Library Cataloguing-in-Publication Data
A catalogue record for this book is available from the British Library

Library of Congress Cataloging in Publication Data
Names: Strebel, Ignaz, 1970– editor. | Silberberger, Jan, 1975– editor.
Title: Architecture competition: project design and the building process / Ignaz Strebel and Jan Silberberger; with contributions from Camille Crossman, Peter Holm Jacobsen, Andreas Kamstrup, Kristian Kreiner, Torsten Schmiedeknecht, Leentje Volker.
Description: New York : Routledge, 2017. |
Series: Design and the built environment |
Includes bibliographical references and index.
Identifiers: LCCN 2016034546 | ISBN 9781472469984 (hardback : alk. paper) | ISBN 9781315567594 (ebook)
Subjects: LCSH: Architecture–Competitions.
Classification: LCC NA2335 .A653 2017 | DDC 720.79–dc23
LC record available at https://lccn.loc.gov/2016034546

ISBN: 9781472469984 (hbk)
ISBN: 9781315567594 (ebk)

Typeset in Sabon
by Out of House Publishing

 Printed and bound by CPI Group (UK) Ltd, Croydon, CR0 4YY

Contents

List of figures	vii
List of tables	viii
Notes on contributors	ix
Editors' acknowledgements	xi
Introduction: unpacking architectural competitions – project design and the building process IGNAZ STREBEL AND JAN SILBERBERGER	1

PART I
Managing the procedure — 29

1 Two geographical logics in architectural competitions — 31
   IGNAZ STREBEL AND JAN SILBERBERGER

2 The competition between creativity and legitimacy — 45
   KRISTIAN KREINER

3 The best of both worlds? Client decision making in architect selection processes — 59
   LEENTJE VOLKER

4 Design in conversation — 78
   AN INTERVIEW WITH MALCOLM READING

PART II
Inside the competition — 85

5 The progressive differentiation of judgement criteria — 87
   JAN SILBERBERGER AND IGNAZ STREBEL

6 Jury board at work: evaluation of architecture and process — 103
   PETER HOLM JACOBSEN AND ANDREAS KAMSTRUP

7 Jury boards as 'risk managers': analysing jury deliberations within architectural competitions against the background of risk management 117
CAMILLE CROSSMAN

8 Competitions beyond spatial specifications 131
AN INTERVIEW WITH DIETMAR EBERLE

PART III
Making the built environment 137

9 The obligatory passage point 139
JAN SILBERBERGER AND IGNAZ STREBEL

10 Architecture as process: OJEU tender and procurement without design 151
TORSTEN SCHMIEDEKNECHT

11 Advanced structural engineering 171
AN INTERVIEW WITH WERNER SOBEK

Index 176

Figures

I.1	Project presentation in Wettbewerbe Aktuell, rubric 'Wettbewerbe weiterverfolgt' [following up on competitions].	18
1.1	Extract from a selected competition preparation document. The furnishing of the space in which the preselection meeting happened is planned in every detail. The layout of the space is presented in a drawing (original scale 1:250) and elements of the set up listed as follows: 54 movable pinboards, 8 corner poles, 50 poles, 22 tables, 28 chairs, 7 back-up chairs.	37
1.2	Registering of competition entries in the Salle Melpomène, Ecole des Beaux-Arts in Paris.	39
4.1	Programmatic and functional requirements taken from the tender brief for the Mumbai City Museum North Wing.	82
5.1	The first page of a two-page overview of the spatial specifications.	88
5.2	Evaluation and assessment criteria for the 'Basel Kunstmuseum, Burghof Extension' competition.	89
5.3	Diagram showing the project perimeter.	91
5.4	Diagram providing essential information on the connecting wing.	92
5.5	'Projekt 2: Neunhundertdreiundvierzig', *Plan#1* and *Plan#3* (original plans presented in the competition measured 118.8 cm × 84.1 cm).	93
5.6	Kunstmuseum Basel, new building.	100
6.1	Timeline.	108
6.2	Visualization of the building in the competition brief.	109
9.1	Project diagram used in interviews.	143
9.2	Rendering representing the winning entry.	145
9.3	Rendering representing the project that obtained the construction permit.	145
10.1	The new Everyman Theatre in Liverpool: principal façade towards Hope Street.	162
10.2	Liverpool Philharmonic Hall: the Entrance Foyer.	162
10.3	Liverpool Philharmonic Hall: the Grand Foyer Bar.	163

Tables

I.1	Phases, subphases and goals of subphases of the building process.	16
3.1	Overview of the case characteristics.	65
3.2	Overview of findings per case in relation to sensemaking processes.	68
9.1	Phases, subphases and goals of subphases of the building process.	140
9.2	Basic facts of the construction projects investigated.	141
9.3	Overview of the various displacements (= mobilization).	147

Contributors

Camille Crossman is a doctoral student in Architecture at the Université de Montréal. She holds a Bachelor and a Master in architectural design. Her research thesis, which is conducted within the framework of the L.E.A.P (Laboratoire d'Étude de l'Architecture Potentielle) under the direction of Prof. Jean-Pierre Chupin, deals with the processes structuring qualitative judgement in the specific context of juries of public architectural competitions.

Peter Holm Jacobsen has conducted empirical studies of architectural competitions and the construction of complex hospital projects in Denmark. He received his PhD in social psychology from Roskilde University in 2014. From 2007 to 2013 he was also affiliated with the Center for Management Studies of the Building Process at Copenhagen Business School. Within the area of architectural competitions he has co-authored a book (in Danish, published in 2013) about the implications of introducing dialogues in architectural competitions in Denmark. He has also co-authored an article about dialogue-based architectural competitions published in the *Scandinavian Journal of Management* (2011). His theoretical research interest includes organization theory, social psychology, situated learning, Science and Technology Studies (STS), project management and economic sociology.

Andreas Kamstrup studied philosophy and organization and is now a PhD Fellow at Copenhagen Business School, Department of Organization, where he is examining how digital technologies influence work. In cooperation with the Danish Architecture Centre he has been examining how a digital platform designed to create innovations within the built environment effects how architectural competitions unfold. Kamstrup has a specific interest in how the expanding digitalization of work processes creates new forms of collaboration and competition. Most recently he has examined this in the technology industries of Silicon Valley through a research stay at Stanford University.

Kristian Kreiner is Professor Emeritus at the Department of Organization at Copenhagen Business School. His research and teaching has covered a wide range of topics in organizational theory, including decision-making, project management, knowledge-intensive work and cross-organizational collaboration. Empirically, he has studied these topics under real, i.e., imperfect circumstances when rational ways of organizing and managing things are not possible. The complexity, uncertainty and ambiguity of construction work – from the early design work to the practical work on site – have inspired the search for other, less rational logics. In relation to architectural competitions, a leading question for him has been to understand what it takes to make incommensurable design entries enter into orderly competition.

Torsten Schmiedeknecht (PhD) is an architect and senior lecturer at the University of Liverpool, where he teaches Design and History & Theory. His research interests are architectural competitions; the representation of architecture in print media; and rationalism in architecture. His research has been published in international journals like *Architectural Research Quarterly*, *Architectural Design* and *Nordic Journal of Architectural Research*. He is the editor of *The Rationalist Reader* (Routledge, 2014, with A. Peckham), *Rationalist Traces* (Wiley, 2007, with A. Peckham, C. Rattray), *An Architect's Guide to Fame* (Architectural Press, 2005, with P. Davies), and *Fame and Architecture* (Wiley, 2001, with J. Chance). He is currently (with A. Peckham) co-editing *Modernism and the Professional Architecture Journal – Reporting, Editing and Reconstructing in Post-War Europe* (Routledge, 2017).

Jan Silberberger is a Post-Doc researcher at ETH Wohnforum – ETH CASE (Centre for Research on Architecture, Society & the Built Environment), Department of Architecture, ETH Zurich, Switzerland. He has studied Architecture and Urban Planning at the University of Stuttgart/Germany and Visual Communication and Fine Arts at the Hochschule für bildende Kuenste in Hamburg, Germany. In 2011 he finished his PhD studies at the University of Fribourg's Geography Unit. His research focuses on decision-making within planning processes.

Ignaz Strebel is a geographer within the research unit ETH Wohnforum – ETH CASE (Centre for Research on Architecture, Society & the Built Environment), Faculty of Architecture, ETH Zurich. He took his PhD at the University of Fribourg in Switzerland in 2003. He was previously a researcher in geography and architecture at the Universities of Glasgow and Edinburgh. His recent work has focused on the urban geographies assembled in work settings, such as, for example, building care, infrastructure maintenance, repair workshops, and the offices of building administration, planning firms and housing administration.

Leentje Volker is an assistant professor in public commissioning at Delft University of Technology, and Secretary of the Dutch Construction Client Forum, a joint initiative of public commissioning clients in the Netherlands aimed at improving the professional level of Dutch construction clients. In 2010 she received her PhD on value judgements and decision making processes of client organizations in design competitions at TU Delft. She has published in several books and journals such as *Design Studies*, *Cities* and *Construction Management & Economics* on topics related to organizing competitions, jury assessments, partner selection and collaboration in construction projects. In her research she combines a psychological perspective on architectural design with management sciences, leading towards innovative insights on the origin of (potential) conflicts in decision making in architecture. The combination of her scientific and managerial activities inspires her to work towards practical solutions of complex issues with a scientific underpinning.

Editors' acknowledgements

The research that lays the groundwork for this volume was carried out at ETH Zurich between 2013 and 2016. The Swiss National Science Foundation gave financial support to this project (no. 100013_143216). The book at hand is the outcome of a collaborative effort involving many colleagues, experts and practitioners in the field. We would like to thank Peter Tränkle, who helped designing the research and collecting data. Various experts have generously shared their knowledge and expertise with us. Special thanks to Monika Kurath, Alain Bovet, Denis Raschpichler, Sofia Paisiou, Bernhard Böhm, Joris Van Wezemael and Jane M. Jacobs, who have provided critique and feedback on various occasions and supported the project from its beginning. We are grateful to Krishna Bharathi, Marko Marskamp and Julio Da Cruz Paulos for feedback during the review process. The volume as a whole would not be comprehensive without the chapters written by established competition researchers and interviewed architects, who agreed to respond to our request to mirror and critically engage with the various topics raised. These parenthesized texts take up, supplement and put our findings in perspective. Furthermore, we would like to extend our thanks to Matthew Carmona and the editorial expertise provided by Ashgate/Routledge, particularly that of Valerie Rose and Sadé Lee.

Introduction
Unpacking architectural competitions – project design and the building process

Ignaz Strebel and Jan Silberberger

Architectural competitions always consist of the same basic constellation: a client poses a task, two or more teams submit their proposals, then an assigned jury board proceeds to evaluate the proposals and selects the most appropriate solution for the task posed. When looking closely at how competitions work, one understands that there is a large spectrum of how this constellation can be assembled. At one end of this spectrum we might find, the open to every potential participant and completely anonymous procedure, in which jury boards evaluate design solutions without knowing who is the architect or the team behind. Within the field of architecture, open procedures are highlighted repeatedly as providing for a range of alternative projects to select from and to produce high quality solutions. Because they are anonymous they are seen as unique opportunities for local and global offices or newcomers and experienced architects to compete on equal terms and to be judged solely on the basis of the design submitted. At the other end of this spectrum, we might find an invitation procedure, in which, for example, a private investor seeks to find an architect he wants to build with. The investor decides to invite a selected number of offices, which are known to him, as potential partners to solve the task posed. At first glance, and compared with the open competition, procedures by invitation do not correspond to criteria of equality and are not necessarily transparent. However, at a second glance, competitions proceeding by invitation are popular with architects and clients alike because they introduce a dialogue between clients and architects from the very beginning.

Having said this, we find ourselves already in the midst of a very exciting debate on what the role and function of competitions in our times should be. For example, fierce defenders of the open competition models would say that one end of this spectrum is governed by coincidence, emotions and informality and at the other reign reliability, trustworthiness and formality. While the followers of the invitation model locate at the one end common sense, mutual trust and sustainable dialogue and at the other a high load of administrative work, randomness in decision making and imposed business partnerships. In this book, we do not want to take a position within this debate and theoretically elaborate on what kind of understanding of architectural design reigns at each end. Instead, we prefer to use the variability displayed in this polarization as a first set of indications that competitions are made of many different parts and notoriously behave erratically. Hence, we propose to attempt investigating how real competition procedures relate and act upon the building process. Conducted in a constructive spirit, this book contributes to the better understanding of – be this in social, technical or artistic terms – effectuality of competition procedures and their deployment in various contexts.

1 Outline of the approach taken

When looking at how architectural competitions are discussed, be this in public, in the trade or non-trade press or in monographs or discussions on competitions have always been centred on specific cases, their set-up and their outcome. What is missing from such discussions is a thorough consideration of the heterogeneity of people, institutions, tools, devices, inscriptions and standards they are made of. Claims to deal more specifically with the complexity of architectural competitions have been made for several years now – however, in most of the cases from a more theoretical stance (Van Wezemael, 2011; Chupin, 2010) or while focusing on the procedure of jury work (Silberberger, 2012; Chupin, 2011; Chupin and Cucuzzella, 2011; Andersson, Kazemian and Rönn, 2010; Spreiregen, 2010; Kreiner, 2006; Spreiregen, 1979). In this book, we want to take this discussion further from a slightly different angle. Our approach is neither theoretically nor professionally motivated, but aims at critically considering the design competition in relation to the building process. We claim that a proper understanding of competitive procedures is only possible when we properly understand what decisions lead to a competition and what effects competitions have on the worlds we live in. In this way, the competition is not a unique, self-contained planning incident, but a device that is embedded in society and the continuous transformation of the built environment.

Our own motivation is less linked to introducing new concepts or new research methods than to contemporary developments within society in general and the businesses of building and procurement in particular that are currently challenging and questioning the tradition of design competitions. We discuss three issues working independently from each other that challenge architectural competitions nowadays:

- the requirement to carry out design competitions in internationally regulated procurement systems;
- the challenge to meet the client's requirements; and
- the task to respond to the growing complexity of both the building process and the built environment.

In the following we will exemplarily summarize what each of these challenges involves; taking information from both literature review and interviews conducted during our research on the post-competition phase of twelve housing competitions in Switzerland (see Table 9.2 in this volume).

1.1 *Public procurement*

The first challenge concerns competitions that are advertised by public authorities and institutions. This is the case when for instance a national or a local government engages in constructing or renovating a public building, such as a school, a town hall, a museum or offices for public administration departments. All public procurement is bound by specific rules. European countries, for instance, follow the so-called Directive 2014/24/EU issued by the European Parliament and the Council, which imposes rules on how to award a tender. Decisive is the fee that the client, for example, the municipality or a department of a municipality, has to pay for the services provided by the architect in the course of designing and realizing the project. Let us substantiate what this means in an example. Switzerland, although not a member of the European Union, signed an earlier version of the above-mentioned EU directive in 1996, which imposes three threshold values in regard to public procurement. For the Swiss city of Zurich, which acquires construction services of an amount of approximately 300 million Swiss Francs per year,[1] this means that every commissioning of architectural services has to be assessed before the city can proceed to the awarding. Regarding the architect's contract, it is important to note that this does not only concern construction work, but every contract awarded by a public authority. The city of Zurich calculates the architect's fee on an estimated construction sum, 10 per cent being a rule of thumb. Regarding the EU directives, if this fee is:

- below 150,000 Swiss Francs, the contract can be directly awarded;
- higher than 150,000 Swiss Francs, the city of Zurich has to organize a competitive procedure;
- between 150,000 and 250,000 Swiss Francs, the city is allowed to conduct an invitation procedure, that is, to directly invite participants to take part in the competition;
- between 250,000 and 350,000 Swiss Francs, the city has to conduct either an open or a restricted procedure;
- higher than 350,000 Swiss Francs, there has to be an international bidding.

Such regulation seems to provide for an environment in favour of the architectural competition, which is seen as a fair awarding procedure. Indeed, this procurement law imposes some of the fundamental principles on which architectural competitions are based: transparency of the procedure, non-discrimination of potential participants and equal information for competitors. However, at a closer look, this marriage of design competitions and procurement law involves serious tensions (Volker, 2011). While the architectural competition is often praised for its experimental nature (Lipstadt, 1989), that is, its ability to produce unforeseen, high-quality solutions by encouraging competing planners and designers to submit out-of-the-box architectural propositions (Silberberger, 2015; 2012), procurement principles call for fairness of conduct and (consequently) for competition briefs with clearly defined specifications and judgement criteria. When looked at closely, procurement law and architectural competition stand in sharp opposition. While the procurement law requests that specifications defined in the competition programme have to be worked off by the competing architects and need to be directly incorporated into their design proposals, design competition in principle works the other way round. The expectation of a design competition is that the architect produces the specifications of the design. This might not be easy to understand, however, the practice of the so-called honourable mention substantiates this specific attribute of the competition. The honourable mention enables the jury to accept that the best solution for the task posed may be in breach with or provide for a significant reinterpretation of the competition brief. The honourable mention stands for the radical character of the competition, it is obviously in opposition to the principles of equal information and transparency, and is therefore problematic for public procurement. From this conflict we learn not only that the design competition is put into question. The tool is controversial not only in architecture, but also in other fields of expertise such as procurement and building law, and has gained the attention of the wider public. While architects use such controversial situations to insist on the fundamental principles of the competition, in this book, we want to consider this conflict as a challenge for design in a democratically structured public market.

1.2 Client requirements

Public procurement regulations obviously concern only those competitions which are advertised by public institutions. The second challenge we want to discuss is the increasingly sophisticated requirements of the client, be it public or private. This is particularly true for private clients and investors, who do not have to follow principles such as transparency, anonymity or equal information. Imagine for example a big corporation that is building its new headquarters in a business district or downtown of a global city. The corporation is willing to organize a competition. However, when considering, for example, whether to carry out a common open competition procedure a series of questions challenge good will and intention (see also the interviews with Dietmar Eberle and Werner Sobek in this volume): Why should the corporation agree to take the risk of not knowing the architect with whom they are going to realize the project before the jury has taken a decision? Why should they agree to involve professionals and experts that are not related to the corporation into the decision making process? Why should they agree to not intervene in how the competition evolves? And why should the corporation agree to judge one hundred projects if they already have a pool of five architects they want to select from?

One specific and crucial feature of the open competition is the anonymity of the submitted project. This means that the office behind the winning design is revealed only after the board of jurors has taken the decision. There are many anecdotes about this issue. From the jury that was sure it had selected a star architect, which in the end was revealed as a no-name (see Crossman in this volume), to the client who had to face a business partner with no local knowledge in regard to building law and industry and whose office and working staff were located in a completely different country, and to the client not warming to the 'externally' imposed architect on a personal level. But these are only some of the problems raised regularly by investors who have to make the decision whether competition is the right way to go. Another issue concerns the ability to participate in the judgement and decision making process. One of the requirements of the open competition is to assemble a jury board consisting of a majority of professional architects and urban designers, meaning that representatives of the client automatically constitute a minority within the jury board. It is not only the giving of power to decide to independent professionals that is at stake. One major doubt comes also with the fact that these experts leave the process directly after the jury sessions and are not available for further consultation. Clients are also aware that once the competition has started, the procedure will go on as agreed until the end without any intermediary assessment that would allow for confirming, reorientating or even abandoning the procedure if necessary. In a similar vein, the client may not see why the competition should proceed to achieve a maximal variety of solutions if many of these solutions are rejected due to not meeting basic requirements. And finally, the client often does not understand why he or she should not limit from the beginning the field of competitors to a few preferred and known offices. Saving time and money is of course on the agenda of every private enterprise. From here, the client has two solutions to choose from: on the one hand, he or she drops the idea that the competition will lead to the expected result. On the other hand, he or she engages in developing a custom-made procedure according to his or her needs. Those who go down the second road, especially, have given new impetus to the competition business. It is at this point where new and adapted competition procedures which involve more dialogue, feedback loops and interventions, have been created in recent years.

1.3 *Complexity of building process and the built environment*

Besides the requests from the public market as well as from investors willing to conduct competitions, a third challenge that competitions face nowadays is raised by those who organize competitions professionally (see, for example, Hoskyn and Müller, 2008). This is that building projects are carried out in increasingly complex environments. Most of them are not realized on green-field sites but intervene into existing built structures and have to resonate with existing urban form and infrastructure. But it is not only the context that poses a challenge for competitions. Clients nowadays often have very detailed understandings of what they are looking for in terms of functionality, usability, durability and cost-efficiency, which reflects increasingly complex affordances, that have to be translated into competition programmes. One consequence of these demands, which often go beyond the realm of design, is a specialization, scientification and managerialization of the building process. At every stage of the building process, organizers have to take into account that values such as energy-efficiency, noise protection, budget for fire safety and maintenance services decide whether a project can be realized or not. What adds to the complexity is that there are not only more standards to follow, but also the set up, implementation and control of such requirements involves new actors that have to be taken care of. This complexity has to be handled and is difficult to control. Single changes can have a major effect on the whole project and put planning reliability into question. In regard to handling the risks and uncertainties that clients, the urban situation and standards pose, competition organizers have to decide if an appropriate response is a very detailed, exhaustive list of specifications for the competing architect or an openly formulated competition brief that allows for interpretation and weighting of the various components involved. This does not necessarily mean that the competition is put into question as a procedure. However, there are significant changes in how entries are judged. This constitutes a challenge for organizers but also has to be considered by competing architects: for example, aspects of sustainability have been granted an increasingly important role in assessing architectural projects. The evaluation of energy efficiency as pushed by sustainability rating systems superposes traditional jury work. While the latter is in its essence a collective process of negotiation governed by design thinking, discussions on energy efficiency tend to be dominated by quantities, which in turn can be determined by individual experts (Cucuzzella, 2013; Chupin and Cucuzzella, 2011).

While competition organizers have to find ways to adequately respond and adapt their practices to the three challenges mentioned, for us, these challenges offer heuristic features not only for the study of competitions as a planning tool, but also as an occasion to work towards a more general understanding of how design and the building process relate to each other. This means that the book at hand is motivated

6 *Ignaz Strebel and Jan Silberberger*

not by normative thinking on competitions, that is, on how to set up, process and bring to an end competitive procedures compliant to rules and convictions. Instead, its aim is to gain a general understanding of how competitions unfold under increasingly complex circumstances and how they contribute to assembling the built environment. This move away from improving and promoting competition procedures is strategically relevant and necessary and does not mean that we do not value the competition as an effective planning tool. However, we think that instead of improving the given standard on the basis of practical experiences and taken-for-granted assumptions that competitions are 'good' planning tools, we suggest taking a detour and looking from a different perspective at what competitions do to society and the built environment.

It is in the nature of every building that from simply looking at it, walking through it or even using it on a daily basis, we cannot discern if it was built by means of a competition or not. The design process is something that remains black-boxed in the building and in recent years many scholars have regarded it as necessary to study and make visible, not only what buildings are, but what buildings are when we look at them as they are 'in the making' (Jacobs et al., 2012; 2007; Guggenheim, 2010; Yaneva, 2009; Latour and Yaneva, 2008; Chapter 9 in this volume, see also below). It is our conviction that research on competitions can contribute to linking realized projects to the design and building process that produced them. It is therefore worth looking at a field of investigation that has in recent years gained shape. What is fascinating about competition research is that despite having a common object of study, it remains until today very heterogeneous. The beginning of competition studies is difficult to locate: competitions have always been accompanied with comments, criticism, reflections and debate, which generated particular knowledge on the procedure. Our story begins in the late seventies when scholarship first appeared that went beyond discussing individual competitions or regulation frameworks and put the focus on the procedure as a general design tool. We want to tell this story before deploying our own approach.

2 The story of competition studies

Every story has a beginning. Finding a starting point within the vast literature and uncountable number of texts on architectural competitions is a difficult task. Competitions themselves produce texts such as competition briefs and jury reports. Competitions in turn are referred to texts such as regulations, norms and best-practice guides. In addition, they are accompanied by myriad articles and comments in the trade and non-trade press. These documents are important tools that have accompanied individual competitions in the past and will accompany them in the future. The often confusing number of documents constitutes a major challenge to every competition researcher. It is difficult to find the right starting point. One of our first observations, when starting to research that vast and permanently growing corpus of documents is what we call here the taken-for-grantedness of the procedure. To say it in technical terms, in most of these texts competitions are seen as procedures that receive an input (as e.g., the design task, the competition entries but also the designated jurors) and that produce an output (as, for example, the winning project or the jury's final report). As diverse as they can be, the majority of these documents focus on these inputs or outputs and from there conclude on what has been done wrong or right while processing the competition. However, very few documents discuss the intrinsic matter of not only how competitions are made (and sometimes unmade) but crucially of what they produce and what they do to the built environment.

In this sense, a welcome number of studies and research initiatives have appeared in recent years and have started conceptualizing competitions, to use Alfred Schütz's terms (1962), in a second-order language,[2] be this from an architectural, historical, anthropological, political or economic point of view. This growing body of literature proves that an appreciation of this design procedure is needed, which is neither normative nor formulated in terms of self-evidence. For us, a good starting point to develop what a critical approach on competitions could be made of are two books published in the 1970s, in the USA and the UK, respectively. The first is Paul Spreiregen's *Design Competitions*, which was published in 1979. The second is Judith Strong's *Participating in Architectural Competitions*, published in 1976. We consider these two key publications as the first systematic attempts to give a comprehensive overview on the internal workings of competitions. We will use these inquiries as starting points for moving on from considering the competition as a ready-made design tool to understanding it as a complex procedure that must be assembled each time anew depending on the specific context and purpose.

From there on, we distinguish four phases in the story of competition studies, meaning that the research perspective has evolved and the field of competition studies has diversified over time. We have named these four phases as follows: 'The good competition' (2.1), 'The social production of architecture' (2.2), 'Competitions in the making' (2.3), and 'Quality through competition' (2.4). We must emphasize that these phases do not replace each other on a timeline but have their individual beginnings and continue co-existing until today. This evolution of competition studies can also be seen as reflected in an on-going series of conferences that brings together the competition research community on a regular basis.[3]

2.1 The good competition

Let us go back to the above-mentioned two books that we use as a starting point for scholarly work on architectural competitions (Spreiregen, 1979; Strong, 1976). These texts were written by two profoundly knowledgeable persons of the USA and the UK competition business respectively, mainly active in the 1970s and also the 1980s. Besides his major publications on urban design, Paul Spreiregen was the first director of Urban Design Programs at the American Institute of Architects and the competition advisor of the Vietnam Veterans Memorial competition (1980–1981) (McLeod, 1989). Prior to publishing her *Guide for Competitors, Promoters and Assessors*, Judith Strong, who was not an architect, had spent seven years in the Public Affairs Department of the Royal Institute of British Architects. As the Competitions Secretary she was responsible for advising on and promoting competitions, and publishing results. Both texts are highly successful in accounting for what was happening inside the competition procedure and provide an overview of the main points of criticism of the general procedure as such, its myths and its mechanisms. Both write from 'inside' the practice and their everyday experience of organizing, assessing or participating in competitions, taking the competition not for granted but as a design process that is shaped by cultural history. Common to both authors is that they see the competition as a process that has evolved over time and is worth defending against those who challenge it. Normative in purpose, their texts not only illuminate what competitions involve, but also what Strong articulates in the introduction to her book as the belief 'that a more flourishing competition system can only help to improve the quality of design, the morale of the profession and the attitude of the public to what is being built around them' (Strong, 1976, p. 1).

What makes Spreiregen's and Strong's accounts so precious is their clear understanding of how competitions work and what role promoters, competitors and assessors have to play. While Spreiregen is more pragmatic, Strong is more systematic. They both provide an overview on types of competition, how the system works in the USA or the UK, what participating architects have to consider in terms of presenting and submitting their projects, and what opportunities they have to get informed before and ask questions to the promoter during the competition. In the case of assessors, Strong is more detailed on who chooses jurors, what their task involves, and how they are committed to the promoter. Both books provide regulations of competitions. While Spreiregen's is an illustrated volume listing a number of projects, including images, Strong discusses competition culture in the UK, Europe and the Commonwealth.

Both texts contrast with the usual documents that the competition business produces. A jury report, an excellent example of a document that presents decisions, does not, however, display the tracks and traces of opinions, debates, changing judgements, practical aspects of organization and jury work. The merit of Spreiregen's book is first of all its comprehensive discussion of the success, problems, pros and cons, functioning and misconception of competition procedures. Spreiregen's declared aim is 'to instate design competitions as useful professional enterprise, serving society' (Spreiregen, 1979, p. 2). Strong's and Spreiregen's accounts involve a good deal of pragmatism, while listing carefully how competitions should work in principle; both also work against general criticism. Taken together they provide a list of arguments that worked against competitions at the time and probably still do. We will list here the seven most important ones taken from both texts:

1. Competitions are costly procedures
2. They take more time
3. The design never gets built
4. There is no dialogue between the client and the architect in the design phase
5. There is risk of choosing an architect that will prove not to be competent to execute the building
6. The buildings labelled as 'good design' are expensive
7. Architectural quality of buildings is given priority over other qualities

Both texts in principle work convincingly against these arguments and therefore, after having opened up the black box and, using technical language, having repaired and strengthened its contents they close the box again convinced that they have left it stronger than it had been before. What we learn from these accounts is that the competition is not 'naturally given' but a practice to produce good design that is agreed upon by a profession and in the wider sense by society. It is clear that the purpose of these accounts was not to critically discuss the ideal model of design competitions. It is interesting to observe how regulating design procedures is always a tightrope walk between proving hard facts and constructing yet another myth.

Particularly in the USA we can observe a continuation of the story of the good competition, which also involves reconsideration and reflection on the rather non-critical 'taken-for-grantedness' that every competition from the start is to be seen as a good thing. Following on from the Vietnam Veterans Memorial competition in 1981 (see below) a number of competitions were organized later in the decade, professional advice was institutionalized and the influential magazine *Competitions* was founded in 1990. This story is told by one of the founding members of the now famous 'Competition Project' (Collyer, 2004) while at the same time elaborating that whether a competition is good or not is not inherent to the competition but lies in the eyes of those who evaluate individual competitions before, during and after they participate.

2.2 The social production of architecture

In spring and summer 1988 the National Academy of Design in New York hosted a mobile exhibition entitled 'The Experimental Tradition'. This exhibition provides the opening for the discussion and study of competitions beyond the desire to provide for 'good' competition procedures and standards. The catalogue of the same title (Lipstadt, 1989), which accompanied the exhibition, was edited by Helene Lipstadt, a cultural historian and anthropologist at MIT School of Architecture and Planning. From her perspective as a social scientist, Lipstadt moves on from praising, idealizing and regulating competitions, to an understanding of competitions as key cultural institutions and activities, which can be studied to better understand the position that architecture holds in society and in local places. Lipstadt is fully aware of the normative and prescriptive tradition of architectural history and architecture that dominated discussions and debates on competitions until then and which established competitions as trendsetting events. She clearly rejects historiographical accounts of architectural competitions (as for example Haan and Haagsma's (1988) overview of international architectural competitions over the last 200 years) that claim that the competition, as it has developed since the French Revolution, is to be regarded as a new form of work organization and therefore has to be seen as a significant driving force of modern architecture. Exhibition and catalogue were based on archival work aimed at documenting non-selected projects in important competitions. Lipstadt was well aware of the shift this would involve when writing in the introduction about what she means by 'Experimental Tradition':

> The notion of 'The Experimental Tradition' should not be taken to signify an unquestioning faith in the benefits of competitions or an affirmation of a historical association of competitions with great style-forming moments of innovation in architecture. Rather, it reflects the nature of the institution of architecture competitions, for these ephemeral events with permanent results are endlessly repeated and always changing. Competitions are battlegrounds of opposing ambitions and antagonistic solutions, giant architecture classrooms with invisible boundaries and, often, open enrolments.
>
> (Lipstadt, 1989, p. 9)

What is remarkable about her work is not only the rejection of the normative appreciation of design competitions, but a demonstration of how the competition came to be historically constructed and aimed at a specific moment in time to discipline uncontrollable and non-desired elements that would impair comparability:

> As the structure of the procedure became regularized in the academies of Italy and France, many of the conditions, techniques, and rules associated with it today were perfected, usually to protect and guarantee that the intrinsic quality of each architectural project could emerge from comparable drawings. The rules governing anonymity, number and type of drawings, their scales, media, and the like were quasi-legal requirements that reduced nonessential differences and thereby permitted the judgment of competence to rest entirely on the essential design. The prescription that all submissions were to be technically similar enforced equivalence and may have also reinforced beliefs about opportunity.
> (Lipstadt, 1989, p. 15)

In this sense the competition re-orders various established relations, which can have positive and negative effects for the architect. First, it separates the project from its author and enforces the autonomy of the project. It is not the architect's competencies that are judged but the project and its qualities. Second, established hierarchies of professionalism are given up. The responsibility for the design is not necessarily put in the hands of experienced architects, giving a chance to young, inexperienced, but potentially excellent architects. And third, the promoter or client is cleared from the responsibility to pay for the work the architect puts into the competition entry.

From this viewpoint we get a different understanding of how architectural competitions came about in time. The competition is no longer seen as a place of progress where new aesthetic styles are generated. For example, in a historical account of architectural competitions in Europe from 1401 to 1927 Bergdoll (1989) deliberately establishes competitions as political devices used in different ways under different power regimes. Although we are not able to elaborate on this study in detail here, we can note that in France, during the Ancien Régime, discussion of architecture and variants of projects was limited to those in power, who called upon members of the *Académy royale d'architecture* to mobilize the necessary expertise. At that time there was already a certain kind of jury board session. However, and this is the crucial point, which project was to be selected was, so to speak, the king's decision. The competition was redefined during the French Revolution and thus became an instrument of democratic and revolutionary architecture. The competition underwent fundamental changes, which were not coherent at all. For example, while in some instances jury boards were elected by participating architects themselves, in other instances they were selected and imposed by the client. It is obvious that through such changes new power structures were established, which would prioritize some social relations over others and particular ways in which the built environment was conceived. As Bergdoll (1989, pp. 35ff.) shows, despite this evolution competitions were sill perceived as corrupt. What is important for a conceptual understanding of the architectural competition, is the fact that historical changes – and in this case the birth of what is called the modern open competition – do not draw back to issues of aesthetic progress within the field of architecture but to requirements of economical, social and cultural contexts.

The Vietnam Veterans Memorial competition (1980–1981) is particularly illustrative regarding the understanding of socially produced architecture. Among others

(Doubek, 2015; Scruggs and Swerdlow, 1985), McLeod (1989) shows clearly that the decision taken was a product of the interplay between people with various goals, established rules, ways of advertising, the field of participants and not least, the dissension between the Vietnam Veterans Memorial Fund and the winning artist over the question of whether the memorial should be an anti-war monument or a war memorial. The competition here is clearly not a selecting machine for the best project, but a platform that provides for contradicting opinions being publicly articulated in the same space. From this viewpoint the functionality of the architectural competition can be seen as having rendered possible the anti-war design of undergraduate student Maya Lin. Competition research has looked at such procedures in detail, aiming, among other things, to show that the final decision is not necessarily a win by one camp over the other. In the case of the Vietnam memorial, for example, the controversy around Maya Lin's project was solved in part by the addition of a war memorial statue to the remembrance park, thus accommodating the views of those originally opposed to the exclusive anti-war character of the monument (see McLeod, 1989).

In a similar vein, historical studies have investigated competitions to understand the influence of public administration on the production of architecture, changes in the relationship between architects and clients, evolving understanding of the city and representations of architecture. However, from competitions we do not only learn about 'social issues' (Becker, 1992). Real time competitions and their products (programmes and jury reports), discussions on regulatory frameworks, and public reactions to the composition of jury boards and decisions taken, reflect procedure and the hopes attached to it. What we get from this history is an account of how a relatively conventionalized and standardized planning tool performs at different moments in space and time. Such accounts confirm also that the architectural competition at its core is a quite robust procedure. Every era and place (socially) composes and practises it in different ways. At the same time, it has outlasted many historical and social changes and until today has stayed a truly young and particularly effective procedure.

Research on architectural competitions is always research on civil society (Malmberg, 2006). Design is grounded in society and politics:

> [A competition] forces designers to adjust their sights and working methods. They need to see themselves as actors in a political system, not floating above it as artists or neutral professionals. Without political skills, they will find their efforts outflanked by those accustomed to acting in the political arena.
> (Sagalyn, 2006, p. 48)

The competition is studied as a platform, which stages those personalities, experts, competences, project authors and publicly active members that make up the history of architecture (Nicolas, 2007). The competition is seen as an excellent heuristic feature in which not only one realized building but many potential, although rejected projects are studied as cultural products (Jaquet, 1995; Koch and Malfroy, 1995). For these scholars the competition is not an object of study but a laboratory in which contemporary society and cities can be observed 'in the making'. As we move to the next phase of our exciting story of competition studies we join a bunch of researchers who have entered this laboratory to explore its ways of working.

2.3 Competitions in the making

The scientific move to study competitions as laboratories in which social groups and material elements come together, which display political constellations, places for design, emerging sites of new socio-technical topologies, knowledge production, and participation is not linear but has been established from 2000 onwards paralleling the two formerly described approaches. The focus on the architectural competition 'in the making', was inspired, on the one hand, by Science and Technology Studies; and on the other hand, by a design methods approach. By 'in the making' we mean looking at the competition not as a ready-made planning technique but as a design tool that is constantly shaped in practice and as it changes relations between the objects, projects and people involved. These studies can be seen as a way of applying Bruno Latour's 'Science in Action' (1987) and the statement that how a technology, a project or an organization is assembled and put to work is best understood by following the actors involved. A substantial body of competition scholarship that adopts this methodology explicitly or implicitly is given in the comprehensive proceedings of the 4th competition conference in Montreal, 2012 (Chupin et al., 2015). We find two research strands worth examining in more detail. The first regards those studies in which researchers, either physically or intellectually, access the offices, meeting rooms, exhibition halls in which those involved in competition work cooperate and talk to each other, as an anthropologist would access a social group as a participant observer. This we call the ethnographic approach. The second strand of research regards studies that use archival data to investigate how competitions and the various human and non-human actors involved relate to each other. This is the archival strand.

To explain what the ethnographic approach means, it is worth going back to a historical study on competitions in the French speaking part of Switzerland. To introduce their historical account, Frey and Kolecek (1995) present in the first part of their edited collection a series of short memos, each a few pages long and written as a personal experience report, on, among other things, the interactions between jurors, the expectations of the client, time management during finalization of a competition entry, and the application of organizational rules to a specific competition. With this series of memos the reader is very nicely introduced to the myriad interactions, constraints and possibilities that competitions offer, and to which a historical account can only make allusion; moreover these memos frame the research field which we scrutinize in this section. In this way, the vast and complex field of what is usually hidden to those who do not directly participate in competitions is opened up and made available for further research.

First we refer to studies that examine how architects work when participating in competitions. More specifically these studies are interested in how architects, in an ideally open and anonymous competition and in a very practical sense, deal with the competition brief and therefore with a client that is not present and not available for communication. This absence of the client is very specific to designing competition projects and is often brought up as one of the main points of criticism against the traditional competition format (Farías, 2013). However, and let us recall this here, to evaluate the good and the ugly or to measure the advantages and disadvantages of a specific competition format is precisely what these ethnographic studies avoid. The question to pursue is rather this: how do architects face the specific dilemma between speculation about the client's aspirations and keeping their feet on the ground? A second problem that is involved in designing for competitions concerns the limited time frame within which architects must hand in their proposal. The difficulty is to deliver on the deadline a project solution that responds to the requirements put forward in the brief in terms of specificities, characteristics, but also economic and technical feasibility. Kreiner (2007) writes of an 'Archimedean point' in the design process that 'allows a consistent design; and it facilitates convincing communication of the design proposal to even the lay members of the panel' (2007, p. 14).

Kreiner points out three ways of reading the competition programme that exist for participants:

> When the brief is read as instructions the challenge is to find solutions that honour the brief without sacrificing other design criteria too much. When read as indications the challenge is to collect additional information about the client and/or the jury to be able to interpret the brief richly and adequately. When read as illustrations the challenge is to make the brief a resource and foundation for the creative exploration of design options.
>
> (Kreiner, 2008, p. 14)

Of these, the first two ways of reading imply that preferences are given in the brief. The third is significantly different. The design preferences are only made public with the submission of the entry. Here it is the participant that shows the client and the jury board what his or her preferences and criteria of good design are (see Kreiner, 2008).

Designers or architects might ask what they can learn from such ethnographic research. It is obvious that such in-depth insight in the work of finalizing a competition entry cannot increase a design team's chances of winning a competition. Still, these studies can be useful for participants themselves before they enter their next competition with regard to organization, such as distribution of tasks, sequencing of work, orientating the design process towards the crafting of the required plans, images, texts, sketches and architectural models, and awareness of what it means to work not with direct business partners but with a brief.

Another field of investigation which has been studied ethnographically, concerns the work of competition juries. In this set of studies the main focus is not on the composition of the board or the professional competences of single members, nor on the dynamic of the group, but on the ways that judgements are made and legitimated. One of the crucial aspects of judgement in jury boards is working against a linear relationship between selection criteria and the decision taken. Jury work has little to do with applying selection criteria to projects but is always a process of retrospective sensemaking: 'the selection of a winner becomes intertwined with the definition of selection criteria. The actual selection helps define selection criteria, as much as the selection criteria help defining the selection of a winner' (Kreiner 2008, pp. 25–26). There is a common understanding within the architectural community that this particular characteristic of architectural judgement is not a weakness of the competition system, but one of its core strengths, allowing learning and producing knowledge. Chupin (2010), using a design methods approach, has fruitfully pushed this observation further. The fundamental assumption here is that there is no architectural project without judgement.

Architectural design emerges through auto-critique, teamwork, teaching and generally speaking exposing the design to peers. The point that Chupin makes about judgement in competitions is that juries do not only learn through judgement but that they actively co-design the winning project. This is a very strong point and probably the statement that until today valorizes jury work most substantially. Elaborating on jury work as collaborative design work Chupin of course does not make allusion to the ways submitted plans and models of single entries would be re-shaped materially by the jury board. What he makes allusion to, is the way a jury board will appropriate a single entry for sharpening its message. This, too, similar to when architects work toward the finalization of the project submission, involves a search for an Archimedean point. Juries work on the consistency of the design's message in regard to convincingly communicating the design proposal in their final reports. The few ethnographies (Cucuzzella 2013; Silberberger, 2012; Kreiner, et al., 2011; Crossman in this volume) that have observed jury deliberations confirm that jury work has a clear effect on the entries judged. They supplement this line of argumentation with records made of judgement instances. Rather than aiming at disqualifying particular decisions taken, the merit of these studies is to make available for analysis a crucial moment of contemporary architectural work that – by definition – has to be carried out behind closed doors. In recent years, architects and architectural societies have highlighted the fact that sustainability standards, building norms and cost calculations have changed the decision making process and that conflicts with the traditional judgement procedure occur. Chupin and Cucuzzella (2011) differentiate between 'decision-making through judgement' and 'decision-making through evaluation or calculation'. While evaluation is based on quantitative assessment and measurement, it is very clear that it differs substantially from judgement, which is based on mutual argumentation, interpretation and critique. With these new research fields new gaps are opening up. We do not, for example, yet understand how evaluation by calculation and measurement as opposed to judgement changes competitions as a whole and in particular project work in architectural offices.

Directly linked to the increasing trust in scientific methods to capture and control the material world that surrounds us, is a complexity of, on the one hand, the built environment and, on the other hand, the building process. It is obvious that in the course of rationalization and technological progress the built environment and the building process must be understood as made of more (in number and in quality) technological elements and as more complicated in essence. However, what we mean by complexity is that buildings, infrastructures, facilities and other built structures have to always be understood as human-made and human-used, which makes them behave in erratic and incomputable ways. Concerning competitions, it is obvious that such complexity does not only affect the very practices of designing and judging within competitions, it concerns the competition business as a whole and each competition as it has to interact with the local specificities.

In this context and with awareness that we do not know much about how competitions evolve in time and space, we would now like to consider several scholars[4] to supplement what we called above the ethnographic approach. The main topic here is the study of how documents, models and related practices can be archived and put in relation. Topics that have been treated with an archival approach concern the question of openness to international participation (Chupin, 2015), the numbers of competitions carried out in various national contexts (Gomes Alves, 2008), the variety of competition procedures applied (Bösch, 2013), and similarities or differences within various data sets, as for instance in regard to a specific task or building type (Katsakou, 2011; Fröhlich, 1995; Jaquet, 1995).

Chupin (2008) has listed four types of record that competitions generate:

- documents that participants receive from the client (e.g., the competition brief, quantitative information, rules, norms);
- the projects of the competitors (e.g., plans, texts, models);
- documents of the judgement process (e.g., meeting protocols, preliminary evaluation reports, final jury reports);
- publications of the results (e.g., final exhibition, press releases, public debate on the result).

Among the various initiatives that have been undertaken to systematically collect competition data (see, for an early example, Frey and Kolecek, 1995) three newer initiatives are worth listing here: the 'Canadian Competitions Catalogue' (www.ccc.umontreal.ca), the German online platform 'Wettbewerbe Aktuell' (www.wettbewerbe-aktuell.de) and the Swiss 'KONKURADO – Web of Design Competitions' (www.konkurado.ch). In relation to these archival initiatives, of importance is the methodological discussion on how to classify, categorize, label and display competition data in an analogical and digital manner (Sorbreira, 2015; Chupin, 2008; Frey, 1998). Methodological discussions also include questions of selection of data, archival work organization and conservation as well as access for professionals and the public. The Swiss platform KONKURADO is an example of an attempt not only to archive data on accomplished and finished procedures, but also to be a platform that is used by competition organizers, jury members and participants at individual stages of the procedure to advertise competitions, exchange documents, ask questions, display results, and so on (Strebel et al., 2015). This platform includes a 'project space', which competition organizers can individually set up according to their preferences and the preferences of the clients for whom they work. If consistently used, KONKURADO will not only be a helpful tool to support the competition process, it also has the advantage of continuously generating and filing 'process-data'.

Within architecture these archives are unique, as their purpose is to classify and store data for a planning procedure and not for a specific building or particular architectural office or institution. Whoever makes the effort to browse one of the above-mentioned catalogues will quickly realize that building culture here is accessed through the memory of the 'un-built'.

To conclude this section: we have distinguished between two research strands that aim at tackling what we call 'competitions in the making'. On the one hand, the ethnographic approach allows us to get inside various stages and work procedures. On the other hand, the archival approach allows reconstruction of the various components of individual or groups of competitions and their state of play at various stages of the procedure. In the following (and final) part of our review of competition research, we will look at how competitions have been conceptualized as one of a series of successive phases and what kind of understanding results from this. We do so in order to introduce our own approach (section 3 below), which aims at understanding the making of competitions as situated in the building process.

16 *Ignaz Strebel and Jan Silberberger*

2.4 *Quality through competition*

Let us go back to the already described situation of the architect at work in his or her studio completing an entry and finalizing a project to be submitted for a competition. We have seen how competition research has highlighted the specificity of this work and particularly the ways she or he incorporates the competition programme in his or her work. Designing for competitions is fundamentally different from architectural work in non-competitive settings. To describe such work settings and particular aspects of the competition is one way to study how competitions work. However, in such instances, we can clearly see how a text that was written at a previous stage of the procedure and by a very different group of actors not only shapes the design process under question but is also re-interpreted if not re-written at the 'next' stage of the procedure. This invites us to rethink what is usually considered to be a linear series of various self-contained phases (see Table I.1) as a far more complex and intertwined course of action. This gives us a different picture not only of the previously discussed studies on competitions in the making but even more importantly on how competitions

Table I.1 Phases, subphases and goals of subphases of the building process.

Phases	Subphases	Goals
Strategic planning	Formulation of needs, solution strategies	Needs, goals and general conditions defined, strategy for solution determined
Preliminary studies	Definition of the project, feasibility study	Procedure and organization defined, project basis defined, feasibility demonstrated
	Selection procedure	Project selected which will best meet the requirements
Project	Preliminary project	Concept and profitability optimized
	Construction project	Project and cost optimized, schedule defined
	Permit-obtaining procedure	Project approved, cost and schedule verified, construction credit granted
Invitation to tender	Invitation to tender, comparison of quotations, application for contract to be awarded	Contract ready for awarding
Implementation	Construction planning	Project ready for implementation
	Implementation	Building structure constructed according to specifications and contract
	Commissioning, completion	Building structure accepted and commissioned, final cost settlement accepted, defects corrected
Management	Operation	Operation ensured and optimized
	Maintenance	Fitness for use and value of the building structure maintained for defined period of time

are generally presented in architectural journals, magazines and in historical accounts as an accumulation of phases of a linear procedure.

Thinking of competition as one stage in a chain of successive phases is the dominant perspective and inherent to the ways competitions are researched, discussed and presented in the architectural field. It is necessary to understand this linear mode of knowledge production in detail before we can introduce our own approach. Let us start by taking a look at the competition magazine *Wettbewerbe aktuell*.[5] This professional journal has a long-standing reputation of being the most complete and, as stated in its title, up to date competition magazine in Germany. It is worth having a look at one of its issues, as it is probably the only magazine in the world which is committed to presenting competitions as sets of representations of stages of the building process. Ideally, this involves photographs of the models of the ranked entries and photographs taken from the realized building. *Wettbewerbe aktuell* is fascinating. The magazine presents the competition procedure as a continuum. This is underlined by a strict use of the same perspective on the building ground, the project and the realized building. The continuity is guaranteed by regular flights undertaken by the magazine's editor-in-chief in a Cessna airplane to particular competition sites in Germany. These flights allow for taking photographs of the realized building from an angle similar to photographs of the architectural models of competition entries and winning projects. This reproduction of the perspective is particularly powerful because it obliterates the long time-distance between the task posed, the projects submitted, the winning entry and the realized building. *Wettbewerbe aktuell* is just one – for our purposes very revealing – example of how competitions are represented in the field of architecture.[6] Such 'before' and 'after' displays are helpful for design professionals to comprehend the evolution of a design. The series of photographs allows for consideration of how the form of a building integrates into a specific site and how it can be seen as reflecting the specifications that were set up by the client in collaboration with members of the jury.

For competition studies, too, such before and after displays are a good starting point – not to reflect upon the evolution of form, but to understand how competition processes unfold. The focus of research is not on how form evolves, but on what happens in practical and organizational terms before, for example, the publication of the programme, the submission of the entries or the selection of the winner.

A particularly interesting entry point to see how competition researchers have stuck to and reproduced the linearity of the building process is Jack Nasar's study (1999) of the use and usability of the Wexner Center for the Arts in Columbus, Ohio. This is an idiosyncratic way of writing about competitions because it focuses not on the competition per se, but on the usability of a building design that was selected through a competition. At first, we find it difficult to add this text to the map of competition research as we deploy it above. Nasar interviewed users of the newly built exhibition building about their satisfaction with its functionality and usability. From his post-occupancy evaluation (POE) he disqualifies the competition as an appropriate process for deciding how a building can and should be used:

> The results of the Wexner POE raise some ethical questions about the competition and the conflict between elite art and popular use. It demonstrates the arrogance and out-of-touch quality of some members of the architectural profession. The POE uncovered few favourable qualities and a host of serious problems with the

Umweltbildungsstätte für das UNESCO Biosphärenreservat Rhön, Markt Oberelsbach
Environment Education Facility for the UNESCO Biosphere Reserve Rhön, Markt Oberelsbach

Architekten/Architects
Neumahr Architekten BDA, Sindelfingen

Mitarbeit
Fabian Eckert · Jennifer Knaak
Jochen Sinnwell

Bauleitung
Architekturbüro Röder, Bad Neustadt/Saale

Landschaftsarchitekt
Neher Landschaftsarchitektur

Fachplaner/Engineers
Statik
Pfefferkorn Ingenieure, Stuttgart
Ingenieurbüro Albus, Bad Neustadt/Saale

HLS
Henne+Walter, Reutlingen
Helfrich Ingenieure, Bad Neustadt/Saale

Elektro
Ingenieurbüro Mück und Schaber, Holzgerlingen
Helfrich Ingenieure, Bad Neustadt/Saale

Bauherr/Client
Markt Oberelsbach

Standortadresse/Location
Auweg 1, 97656 Oberelsbach

Projektdaten/Technical Data
Wettbewerbsdokumentation siehe Heft 2/2010
Platzierung des Wettbewerbsentwurfes 2. Preis
Bauzeit September 2010 bis Dezember 2011
Bruttogeschossfläche (BGF)   2.360 m²
Bruttorauminhalt (BRI)   8.312 m³

Fotos/Photographs
Stephan Neumahr, Sindelfingen

Luftfoto/Aerial Photo
wa wettbewerbe aktuell

Kommentar der Architekten
Das Biosphärenreservat Rhön wirbt mit dem poetischen, wie gleichermaßen realistischen Slogan „Land der offenen Fernen". Diese „offenen Fernen" sollen von jedem Zimmer erlebbar sein. Aus diesem Grund wurde ein dreigeschossiger Baukörper realisiert: zwei Zimmergeschosse mit traumhafter Fernsicht liegen auf einem Erdgeschoss mit Gemeinschaftseinrichtungen. Eine Freitreppe führt unter dem Ring hindurch zum Haupteingang in ein großzügiges, lichtdurchflutetes Foyer. Dort erschließt ein übersichtliches, kleines Rundlaufsystem um einen begrünten Innenhof sämtliche Gemeinschaftsbereiche: die Verwaltung, die Seminarräume, die Küche und die Cafeteria. Der östlich gelegene Wirtschaftsbereich wird separat von außen erschlossen. Dieses Geschoß liegt auf derselben Ebene wie die Sporthalle und ermöglicht durch einen zweiten Zugang sowohl den Besuchern der Umweltbildungseinrichtung wie Benutzern der Sporthalle eine zweiseitige „durchlässige" Verbindung. Die Seminarräume im Osten und Norden orientieren sich direkt zum Außenbereich, die Cafeteria nach Süden zum Eingangsbereich.
Eine zweiläufige Treppe und ein Aufzug erschließen die beiden Unterkunftsgeschosse, deren Zimmer sich nach Osten, Süden und Westen orientieren. Von den insgesamt 36 Doppelzimmern sind 4 als schaltbare

Figure I.1 Project presentation in Wettbewerbe Aktuell, rubric 'Wettbewerbe weiterverfolgt' [following up on competitions].
Note: The three images show the development of the winning entry of the competition for the Environment Facility for the UNESCO Biosphere Reserve Röhn in Markt Oberleisbach, Germany. On top left the model of the winning entry and bottom the building as realized (both photographs by Stephan Neumahr, the winning architect). The areal photograph on top right was taken by Wettbewerbe Aktuell.
Source: Wettbewerbe Aktuell 3/2012, p. 89. Architect: Neumahr Architekten BDA, Sindelfingen, Germany.

design. It identified design problems disrupting secondary or support functions, such as the bookstore or café, as well as design problems disrupting the main function of the facility – the display or performance arts. [...] For a public entity like the Wexner Center, a higher standard applies. In a democratic society, the process should lead to solutions that benefit the public – durable solutions that work for and delight the client, users, and the public.

(Nasar, 1999, pp. 143–144)

It seems that Nasar successfully puts into perspective the problems between design process and the operability and usability of a building. In his framework the competition is situated, not just as one step in the building process, but as the decisive phase having the major impact on what the final, occupied building will be. We cannot deny that Nasar has a point as he alludes to how the competition emanates beyond the decision taken by the jury board and its conclusion. The assumption that competitions produce 'elite art' and not 'popular use' might be exaggerated; however, it brings a particular categorization of the relationship between design and the building process to the fore. What we think is that working with such polarization, that is, comparing two successive, yet not connected phases of the building process and establishing non-coherence between them, is only possible with an agenda of the linearity of the building process operating in the background, as we have elaborated above. This is particularly problematic for complex systems such as the built environment and the building process, which can behave erratically and do not operate in a linear mode. Comparing users' needs to deficits of the philosophy of the designers involved excludes an understanding of how certain qualities circulate within the building process and others do not. How is it that a whole jury board is not able and obliged to imagine what users' needs are? Did they discuss users' needs, but maybe the wrong ones? If users' needs are discussed, how do they materialize in the building or why do they not make it into the jury report? If users are not involved in the design process, are they consulted and considered during furnishment? Such questions suggest not only that more study of the details of the building process are required, but that it is necessary to overcome the linear thinking involved and to be aware that once defined, qualities (costs, atmospheres, usability, energy efficiency, etc.) at every stage of the building process can intensify, transform, disappear or new qualities can appear unexpectedly. To understand the competition as situated in the building process therefore means to elaborate on a manner of investigating what those involved have to do to get from a 'before' to an 'after' condition.

3 Project design and the building process

This book now intends to pursue this new programme. Its main goal is to add to this understanding not only by studying jury board meetings, programme writing and project design, but crucially by assuming that what is done in practice in these various places is embedded within and shapes a larger process, this is to say that competitions are always made and brought forward in relation to the building process. The aim of the book at hand is then to better understand architectural competitions as circulation devices of architectural and other qualities of the built environment. The question to be asked then is how in each of the mentioned places such qualities are introduced if absent, or reconsidered and rejected if already there; how they are transported to the next step and how they materialize in a building, infrastructure or any other part of the built environment. We also want to ask how such qualities are reformulated in various instances of competitions and how sometimes they become distorted, misrepresented or abandoned. To this end we focus on how various actors, that respond to the immediate task posed and deal with the situation in which a task has to be accomplished, retrospectively and prospectively orientate towards what happened in the past and what might happen in the future.

We have already said that we could use a terminology which highlights that the qualities defined and formulated in the building process should be understood as circulating and moving within the building process. The terminology of this conceptual approach has been taken from Actor-Network Theory (ANT); for our purposes the original naming of ANT as the Sociology of Translation is also useful. The latter name emphasizes the importance understanding the building process not as made of successfully connected and mutually responding phases, but as a flow of transformations. Bruno Latour's study (1993) in the sociology of science on how soil scientists mobilize data samples of the Brazilian rainforest from their original place in Brazil, to their laboratories in France and further into written scientific reports is a good starting point. For our purposes such chains of translations are useful, as the scientific work and fact finding is conceptualized as a process in which reference circulates 'like electricity in a wire'. It is only if one can travel back and forth on this chain of translation that the scientific facts are stable and true. Nasar certainly does not formalize the building process in similar ways. For him there is only a beginning (elite art) and the end (popular use), which do not match each other. If, alternatively, the building process were thought of as a chain of translation, a different understanding of the process would be possible. Take such diverse qualities as the atmosphere of a building; its efficiency in terms of energy, aesthetics, usability; its relation to the urban environment in which it is located, and think about how it is that some of these qualities first of all enter and then are able to reach the very end of the translation chain and others are abandoned or distorted during the process. In analogy to Latour's rain forest investigation, we could speak about quality that circulates in the building process like electricity in a wire.

The analogy between the making of science and the making of a building is certainly very tempting. Such analogical thinking is, however, problematic, when we think that every new or existing association between those people and technologies involved in the building process (for example individuals, organizations, tools, objects, documents, technologies, materials, vehicles) have different effects on how the process progresses. Building processes integrate various trajectories with different goals and they are kept going by the ever changing associations that we have already mentioned. To introduce a non-linear understanding of the building process, various authors have suggested speaking of infrastructure, the city or the built environment as an assemblage (Farías, 2011; McFarlane, 2011; Farìas and Bender, 2010). In this volume we suggest using 'assemblage' as way to understand how the building process is shaped, by whom and what is involved. It also helps in understanding why some key elements of the building process are kept circulating longer than others. Assemblage helps us to accommodate the architectural competition's particular context and challenges that we have sketched above, such as the conflict between design and international rules of public procurement, the difficulties of matching clients' requirements, or the complexification of the built environment. To better understand these relations and also questions such as why some buildings do not fit their usage, we aim at using those assemblage thinkers within the field of urban studies who are compatible with the political concerns of the urban studies tradition (for example, Jacobs, 2011). From such a perspective we then can ask why in specific competitions particular forms of processes or urbanisms repress others and how over time they produce and maintain asymmetrical relations of power and unequal availability of resources (see, for example, McFarlane, 2011).

This conceptual thinking would certainly help towards understanding the architectural competition as an assemblage of heterogeneous components on its own or as an assemblage that is related to other assemblages such as the public market, the institution behind the client or the built environment. However, what we intend to add to this conceptual framework is a methodological move in that it is not our intention to conceptualize the competition as an assemblage, but to take a step back and to empirically investigate how competitions as specific instances or situations assemble the building process. We therefore want to put forward the competition as a situated activity (Suchman, 1987; 2007) to contrast an understanding of the competition as something that is regulated by norms and carried out according to agreed upon instructions. When looking closer at competitions one must agree that the relationship between plan (norms, standards, instructions) and action has to be reconfigured. A striking example taken from Lucy Suchman's book *Plans and Situated Actions* (1987, new edition, 2007) explains this very clearly. Imagine a canoeist sitting on a riverbank and studying the waterfall he intends to pass in his canoe. 'I'll get as far over to the left as possible, try to make it between those two large rocks, then backferry hard to the right to make it around that next bunch' (Suchman, 2007, p. 72). Suchman emphasizes that such a plan does not actually describe how canoe and canoeist will get through the difficult passage. To respond to unexpected water currents and to navigate the narrow boat through difficult water the canoeist will fall back not on his original plan but on available skills and abilities. The plan itself is at most either an instrument of orientation helping the canoeist *before the action takes place* – for instance, find a good starting position from where best to tackle a difficult part of the river – or the plan helps him, *after the action has taken place*, to evaluate and valorize his skills, reactions, faults and movements, and whether they were successful or not. The waterfall is an illustrative example that can be used to understand the a priori and a posteriori function of the plan. This is also useful for the competition and probably every design process since the illustration suggests a clear detachment of instructions from the course of action. Suchman does not mean that instructions can play absolutely no role while an action takes place. Plans, for example, can be configured and reconfigured ad hoc as people pursue their tasks and make their next moves related to the situation under way. But the plan remains a representation of an action and not the reason for the action itself. 'The fact that we can perform a post-hoc analysis of situated action that will make it appear to have followed a rational plan says more about the nature of our analysis than it does about our situated action' (Suchman 2007, pp. 72–73).

While in the previous section of this introduction we were critical about comparing states of 'before' and 'after', we now can illustrate how we would reconsider, for example, decision making in jury deliberations. The competition brief, which is clearly a kind of plan of what should happen in such a session, gives the jury board an orientation on how to navigate through the many entries to be judged, but it does not master the jury board in its actions and take it through the problems it has to solve. Similarly, the final report is not a report of what happened inside the jury board meeting, but a text intentionally designed for publication. This means that inside the jury board meeting there is simultaneously an orientation to observe a priori formulations of the design problem and an orientation to provide for a posteriori accounts of what the jury board has done. Moreover, it is not only in jury board settings that such orientations can be observed, but also during every process through which buildings are made. This is the purpose of this book: to look at such retrospective and prospective orientation of the building process related to the competition.

4 Organization of the book

This book is divided into three parts. Each part will look at one specific aspect of how unfolding competitions relate to the building process. The first part looks at how the procedure manages and sometimes struggles to stick with the competition brief and regulation. The second part focuses on the judgement process and how it can be variously framed. The third part then displays how competitions are orientated towards realization of the building. Each part of the book is structured in similar ways. Each opens with one study from our own research exploring how competition work is shaped and differently orientated at the stages mentioned. These contributions will then be supplemented by two chapters in Parts I and II and one chapter in Part III written by competition researchers, who tackle similar themes and elaborate from their own viewpoint on the ways competitions integrate and sometimes disintegrate the rules, objects, people, organizations and time schedules of which they are made. Each part will end with an interview of a practitioner and expert involved at various stages of competition driven building processes.

4.1 *Managing the procedure*

This structure has special relevance to the first group of four texts gathered in Part I under the title 'Managing the procedure', which is inspired by research on how particular ways of organizing and setting up competitions are crucial for every competition to proceed appropriately. Besides explaining what competitions are made of, the chapters also ask what happens if organizational aspects are considered as a hindrance to selection of the 'best' projects.

The introductory chapter focuses on two organizational aspects of competitions. It is devoted to describing the geographical and spatial logics that govern competitions and how they are organized and carried through. The first such logic is how the internationality of participation fields is structured through competition programmes and the second is how equal treatment is supported by the ways jury board meetings are organized in time and space. The two following chapters and the interview pursue the quest of how organizational work, before decision making takes place, influences and shapes the procedure as a whole.

Kristian Kreiner examines how the competition for the new Maritime Museum in Denmark struggled between legitimacy – following the rules – and creativity – choosing the best project. The case shows the dilemma the jury board was in to accommodate a best project that definitively did not follow the rules. During the jury process, rules, which were made before the competition, were re-made to allow a particular project to win, during the competition. While one could say that such retrofitting violates every ethical standard on how competitions are run, Kreiner generalizes this aspect, showing that retrofitting rules is an integral and essential part of decision making and fundamental for every act of creativity.

Leentje Volker looks at a similar problem as she investigates tensions and conflicts of decision making processes in tenders for architectural services. Her case studies are a school with sports facilities, a city hall with library, a provincial government office building and a faculty building. For each of these cases the author describes the ways justification of decision tasks, decision processes and decisions made involves a great deal of sensemaking during and after the competition. The merit of this chapter is that it shows how this process is shaped by the conflict between the rationality of procurement law and the world of architectural design.

Part I concludes in conversation with Malcolm Reading, head of MRC, a company with experience in organizing competitions in the UK and worldwide. Following an ethics of open management and an awareness of how best to trigger the potential of creativity, we learn how selection of jury board members, use of visualizations in competition briefs, and well crafted relationships between client and architect become an integral part of competition organization.

4.2 Inside the competition

While Part I discusses how organizational aspects are reflected in and have sometimes to be re-specified during the competition, the chapters in Part II, 'Inside the competition', focus on jury-based decision making within architectural competitions and elaborate how this specific practice is simultaneously both retrospectively and prospectively orientated towards the eventual construction process.

The opening chapter discusses jury work in the competition for the extension of the Kunstmuseum Basel that took place in 2009. The focus here is on how during the jury process specific projects are rejected and later reintroduced in the pool of potential candidates. The chapter follows one of the entries, elaborating how it was rejected in the first instance and later reintroduced into the selection process. In particular the chapter undermines the often hastily made assumption that jury boards will consider one single project to discuss whether it fills the criteria formulated in the competition brief, and then proceed to selection or rejection.

Peter Holm Jacobsen uses a competition related to the redevelopment of the former Carlsberg brewery site in Copenhagen. This competition built on dialogue: projects were presented by the architect teams to a panel consisting of representatives of the client, advisors, future users and the jury board. The chapter reveals how the discussions between panel and architect teams, and within the panel produced social interactions and revealed a potential for participation in the selection process. It also highlights how such procedures produce dilemmas, particularly for the architects, who realized that with the dialogue judgement criteria underwent changes and suddenly were significantly different from those that were formulated in the original competition brief, which the architects incorporated during work on the project.

Camille Crossman analyses how during jury board deliberations, members show awareness that they do not only judge *actual* properties of individual projects, but have to assess the potential risk involved in each project and in particular the capability of the potential architect to execute the project if selected. Using data from an

24 *Ignaz Strebel and Jan Silberberger*

ethnographic study on the work of jury board members in four Canadian architectural competitions, the chapter focuses on the importance of speculations made about, for example, the age of the team and the experience it might have.

In the interview that concludes this section, architect Dietmar Eberle, from his viewpoint as a competition participant and jury board member, develops a critical perspective towards current trends in the competition business. In the spotlight here stands, among other things, the dilemma of the open competition, which focuses on projects and not on who would be the client's potential future partner architect. Eberle discusses how competition organizers have responded, perhaps unconsciously, to the uncertainties that come with the open competition by introducing masses of criteria and specifications listed in competition programmes. Such an overload, according to the architect, leads to mediocre design and is in no way a warrant for good quality.

4.3 *Making the built environment*

Following from this, Part III of the volume, 'Making the built environment', deals with the question of how competitions always orientate towards what comes next. In particular, the chapters will elaborate on how qualities and requirements often defined as mandatory during the competition are sometimes omitted during the subsequent project phase.

The introductory chapter uses the case of the cooperative housing project Kalkbreite that was built from 2012 to 2014 on a brown field within the city of Zurich in Switzerland. The chapter does two things: first, it conceptualizes the competition that was carried out to select a convenient project as a threshold that had to be passed by the people and institutions involved: the department of urban planning, zoning codes, architectural offices, potential business tenants, future inhabitants. Second, it shows how the associations established during competition between these elements had to be carefully reconsidered and reorganized as a new business tenant put at least parts of the selected and to be built project into question after the competition. Here we learn something about the sustainability and stability of decisions produced in competitions.

Torsten Schmiedeknecht focuses on the use of Official Journal of European Union (OJEU) procedures for two construction projects in the city of Liverpool: the new Everyman Theatre completed in 2013 and the refurbishment and the new extension of the Philharmonic Hall in 2015. The competition is revealed in this chapter as process-orientated as opposed to product-orientated. The chapter reveals how discussion within the jury boards was very much orientated to what comes after the competition as it focused on the potential of the selected architect to develop, fine-tune and not least build the project under question.

This part too closes in conversational style with an interview with architect and structural engineer Werner Sobek, who from his background as a structural engineer, at first glance, might be considered as somebody that joins work on the project after the competition has been decided. From this perspective, Sobek reflects on his experience of coming rather late into the building process, yet he also tells us about the added value that advanced intervention by the structural engineer can have on the project and its potential to productively contribute to the making of a competition entry.

Notes

1 Immobilien Stadt Zürich, www.stadt-zuerich.ch/immo (last retrieved on 26 August 2015).
2 For Schütz, social science concepts are of second order because they are produced from empirical investigation into actors' common-sense meaning or their first-order constructs, respectively. It is by building on the constructions and of the actors in the field that the researcher links the everyday world with the world of theory.
3 The first conference took place in Stockholm in 2008 (see Rönn, Kazemian and Andersson, 2010) followed by a session on competitions at the Construction Matters conference in Copenhagen in May 2010; next conferences took place in Montreal in 2012 (see Chupin, Cucuzzella and Helal, 2015); Helsinki, also in 2012; Delft in 2014 (see Volker and Manzoni, 2014); and Leeds in 2016.
4 A list of contemporary initiatives is given below.
5 www.wettbewerbe-aktuell.de
6 See for another example the quarterly magazine *Hochparterre Wettbewerbe*, www.hochparterre.ch/publikationen/hochparterrewettbewerbe (last retrieved on 25 April 2016).

References

Andersson, J. E., Kazemian, R., and Rönn, M. (2010). The Architectural Competition: Research Inquiries and Experiences. In M. Rönn, R. Kazemian and J. E. Andersson (Eds), *The Architectural Competition: Research Inquiries and Experiences*. Stockholm: Axl Books.
Becker, H. (1992). *Geschichte der Architektur- und Städtebauwettbewerbe*. Stuttgart: Kohlhammer.
Bergdoll, B. (1989). Competing in the Academy and the Marketplace: European Architecture Competitions, 1401–1927. In H. Lipstadt (Ed.), *The Experimental Tradition*. New York: Princeton Architectural Press.
Bösch, I. (2013). Das Jahr des Offenen Wettbewerbs. Hochparterre [Online]. Available: www.hochparterre.ch/nachrichten/wettbewerbe/blog/post/detail/das-jahr-des-offenen-wettbewerbs/1363012487/ (last retrieved on 25 April 2016).
Chupin, J.-P. (2008). Documenting Competitions, Contributing to Research, Archiving Events. In D. Peyceré, F. Wierre and C. Koch (Eds), *Architecture and Digital Archives. Architecture in the Digital Age: A Question of Memory*. Gollion: Infolio.
Chupin, J.-P. (2010). Conflicts of analogical interpretation in architectural judgement: A research program on competitions. Copenhagen Conference on Research in Architectural Competitions.
Chupin, J.-P. (2011). Judgement by design: Towards a model for studying and improving the competition process in architecture and urban design. *Scandinavian Journal of Management*, 27, pp. 173–184.
Chupin, J.-P. (2015). The Candadian Competitions Catalogue: Digital Libraries of Projects as Collective Legacy. In J.-P. Chupin, C. Cucuzzella and B. Helal (Eds), *Architecture Competitions and the Production of Culture, Quality and Knowledge*. Montreal: Potential Architecture Books.
Chupin, J.-P., and Cucuzzella, C. (2011). Environmental standards and judgment processes in competitions for public buildings. *Geographica Helvetica*, 66, pp. 13–23.
Chupin, J.-P., Cucuzzella, C., and Helal, B. (Eds) (2015). *Architecture Competitions and the Production of Culture, Quality and Knowledge*. Montreal: Potential Architecture Books.
Collyer, S. G. (2004). *Competitions: Competing Globally in Architecture Competitions*. Chichester: Wiley-Academy.
Cucuzzella, C. (2013). When the Narrative of Environmental Certifications Replaces the Debate on Quality. In C. Mager, L. Matthey, D. Gaillard and H. Gallezot (Eds), *Faire des histoires? Du récit d'urbanisme à l'urbanisme fonctionel: Faire la ville à l'heure de la société du spectacle*. Geneva: Fondation Braillard Architectes.
Doubek, R. W. (2015). *Creating the Vietnam Veterans Memorial: The Inside Story*. Jefferson, NC: McFarland.

Farías, I. (2011). The politics of urban assemblages. *City*, 15, pp. 365–374.
Farías, I. (2013). Heteronomie und Notwendigkeit. Wie Architekt/innen Wettbewerbsbeiträge entwickeln. In Seminar für Europäische Ethnologie/Volkskunde der Universität Kiel (Ed.), *Kulturen des Wettbewerbs*. Münster, New York: Waxmann.
Farías, I., and Bender, T. (2010). *Urban Assemblages*. London: Routledge.
Frey, P. (1998). *Histoire et archives architecturales: éléments méthodologiques et informatiques: le fonds Alphonse Laverrière aux Archives de la Construction Moderne*. Lausanne: École Polytechnique Fédérale de Lausanne.
Frey, P., and Kolecek, I. (Eds) (1995). *Concours d'architecture et d'urbanisme en Suisse romande. Histoire et actualité*. Lausanne: Editions Payot.
Fröhlich, M. (1995). Edilité publique féedérale: la Poste, 1885–1902. In P. A. Frey and I. Kolecek (Eds), *Concours d'Architecture et d'Urbanisme en Suisse Romande*. Lausanne: Payot.
Gomes Alves, L. (2008). *L'archivage numérique des projets 'Europan' comme situation d'analyse scientifique du concours d'idées en architecture*. (Mémoire de maîtrise effectué sous la direction du professeur Jean-Pierre Chupin). Montreal: Université de Montréal.
Guggenheim, M. (2010). Mutable Immobiles: Building Conversion as a Problem of Quasi-Technologies. In I. Farías and T. Bender (Eds), *Urban Assemblages*. London: Routledge.
Haan, H. D., and Haagsma, I. (Eds) (1988). *Architects in Competition: International Architectural Competitions of the Last Two Hundred Years*. New York: Thames and Hudson.
Hoskyn, J., and Müller, U. (2008). Qualität Dank Wettbewerb. In Amt für Hochbauten (Ed.), *Das Amt für Hochbauten 1997–2007*. Zurich: Verlag Neue Zürcher Zeitung.
Jacobs, J. M. (2011). Urban geographies I: still thinking cities relationally. *Progress in Human Geography*, 36, pp. 412–422.
Jacobs, J. M., Cairns, S., and Strebel, I. (2007). 'A tall storey … but, a fact just the same': the Red Road highrise as a black box. *Urban Studies (Special Issue on 'Supertall Living')*, 43, pp. 609–629.
Jacobs, J. M., Cairns, S., and Strebel, I. (2012). Doing building work: methods at the interface of geography and architecture. *Geographical Research*, 50, pp. 126–140.
Jaquet, M. (1995). Dessine-moi une école! In P. A. Frey and I. Kolecek (Eds), *Concours d'Architecture et d'Urbanisme en Suisse Romande*. Lausanne: Payot.
Katsakou, A. (2011). Recent architectural competitions for collective housing in Switzerland: impact of this framework on architectural conception and innovation. Lausanne: EPFL.
Koch, M., and Malfroy, S. (1995). Vers un urbanisme des experts. In P. A. Frey and I. Kolecek (Eds), *Concours d'Architecture et d'Urbanisme en Suisse Romande*. Lausanne: Payot.
Kreiner, K. (2006). *Architectural Competitions: A Case-Study*. Copenhagen: Center for Management Studies of the Building Process, Copenhagen Business School.
Kreiner, K. (2007). Constructing the client in architectural competition: an ethnographic study of revealed strategies. Paper presented at the 23rd EGOS Colloquium, Vienna.
Kreiner, K. (2008). Architectural competitions: empirical observations and strategic implications for architectural firms. Paper presented at the conference, Architectural Competitions. Stockholm Department of Organization, Copenhagen Business School.
Kreiner, K., Jacobsen, P. H., and Jensen, D. T. (2011). Dialogues and the problems of knowing: reinventing the architectural competition. *Scandinavian Journal of Management*, 27, pp. 160–166.
Latour, B. (1987). *Science in Action: How to Follow Scientists and Engineers Through Society*. Milton Keynes, UK: Open University Press.
Latour, B. (1993). Le 'pédofil' de Boa Vista – montage photo-philosophique. *Petites leçons de sociologie des sciences*. Paris: Seuil.
Latour, B., and Yaneva, A. (2008). Give Me a Gun and I Will Make all Buildings Move: An ANT's View of Architecture. In R. Geiser (Ed.), *Explorations in Architecture: Teaching, Design, Research*. Basel: Birkhäuser.
Lipstadt, H. (Ed.) (1989). *The Experimental Tradition*. New York: Princeton Architectural Press.

Malmberg, C. (Ed.) (2006). *The Politics of Design: Competitions for Public Projects*. Princeton: The Trustees of Princeton University.
Mcfarlane, C. (2011). The city as assemblage: dwelling and urban space. *Environment and Planning D: Society and Space*, 29, pp. 649–671.
Mcleod, M. (1989). The Battle for the Monument: The Vietnam Veterans Memorial. In H. Lipstadt (Ed.), *The Experimental Tradition*. New York: Princeton Architectural Press.
Nasar, J. L. (1999). *Design by Competition: Making Design Competition Work*. Cambridge: Cambridge University Press.
Nicolas, A. (2007). *L'apogée des concours internationaux d'architecture: L'action de l'UIA 1948–1975*. Paris: Picard.
Rönn, M., Kazemian, R., and Andersson, J. E. (Eds) (2010). *The Architectural Competition: Research Inquiries and Experiences*. Stockholm: Axl Books.
Sagalyn, L. B. (2006). The Political Fabric of Design Competitions. In C. Malmberg (Ed.), *The Politics of Design: Competitions for Public Projects*. Princeton: The Trustees of Princeton University.
Schuetz, A. (1962). *Collected Papers I: The Problem of Social Reality*. The Hague: Nijhoff.
Scruggs, J. C., and Swerdlow, J. L. (1985). *To Heal a Nation: The Vietnam Veterans Memorial*. New York: Harper and Row.
Silberberger, J. (2012). Jury sessions as non-trivial machines: a procedural analysis. *Journal of Design Research*, 10, pp. 258–268.
Silberberger, J. (2015). The Honourable Mention. In J.-P. Chupin, C. Cucuzzella and B. Helal (Eds), *Architecture Competitions and the Production of Culture, Quality and Knowledge*. Montreal: Potential Architecture Books.
Sorbreira, F. (2015). Design Competitions in Brazil: Building a [Digital] Culture for Architectural Quality. In J.-P. Chupin, C. Cucuzzella and B. Helal (Eds), *Architecture Competitions and the Production of Culture, Quality and Knowledge*. Montreal: Potential Architecture Books.
Spreiregen, P. (1979). *Design Competitions*. New York: McGraw Hill.
Spreiregen, P. (2010). The Vietnam Veterans Memorial Design Competition Washington DC, 1980–1981. In M. Rönn, R. Kazemian and J. E. Andersson (Eds), *The Architectural Competition: Research Inquiries and Experiences*. Stockholm: Axl Books.
Strebel, I., Silberberger, J., and Raschpichler, D. (2015). Capturing Competition Data: Involving Stakeholders in a Swiss Competiton Database. In J.-P. Chupin, C. Cucuzzella and B. Helal (Eds), *Architecture Competitions and the Production of Culture, Quality and Knowledge*. Montreal: Potential Architecture Books.
Strong, J. (1976). *Participating in Architectural Competitions: A Guide for Competitors, Promoters and Assessors*. London: The Architectural Press.
Suchman, L. (1987). *Plans and Situated Actions: The Problem of Human–Machine Communication*. Cambridge: Cambridge University Press.
Suchman, L. (2007). *Human–Machine Reconfigurations: Plans and Situated Actions*, 2nd Edition. Cambridge and New York: Cambridge University Press.
Van Wezemael, J. (2011). Research on architectural competitions: towards a theory of jury-based decision-making. *Scandinavian Journal of Management*, 27, pp. 157–159.
Volker, L. (2011). *Deciding about Design Quality: Value Judgements and Decision Making in the Selection of Architects by Public Clients under European Tendering Regulations*. Leiden: Sidestone Press.
Volker, L., and Manzoni, B. (2014). Proceedings of the 5th International Conference on Competitions. Delft University of Technology.
Yaneva, A. (2009). *The Making of a Building: A Pragmatist Approach to Architecture*. Bern: Peter Lang.

Part I
Managing the procedure

1 Two geographical logics in architectural competitions

Ignaz Strebel and Jan Silberberger

Introduction

This chapter focuses on the role of architectural competitions in contemporary, and often global, planning and design procedures. While we will argue that these competitions shape both architectural practice and urban form, we will explicitly use a geographical lens, one interested in and attentive to the spatial logics and relations of the practices that come together in competitions and the consequent influence they have in shaping actual urban morphologies. As we will see, buildings that arise from architectural proposals generated in and through competition processes are mediated and assembled by the competition process itself. This reminds us of the collaborative nature of architectural agency. Within the profession of architecture the competition has attracted many, largely unchallenged plaudits (de Jong and Mattie, 1994a; 1994b; Spreiregen, 1979). In professional associations, for example, the competition is regarded as an 'engine of progress standing for quality and innovation'[1] or as 'contribut[ing] outstandingly to the creation of quality architecture'.[2] Despite the apparent consensus that competitions are an excellent way to select the right design solution for a specific architectural brief, there is relatively little known about how they work. How are competitions, international or others, bound to local contexts? Who participates in them and why? How do they transport ideas and concepts? What, exactly, is their innovation potential? Do competitions really ensure the best designs and solutions are selected for appropriate places? In the introduction to this volume, we have already mentioned that the existing scholarly work on competitions is largely uncritical and even idealizes the power of the process to improve the quality of the architectural designs and maximize the outcomes for the soliciting clients and/or context. In this literature a good deal of attention has been given to how competitions have been seminal processes for innovations in architectural style (de Jong and Mattie, 1994a; 1994b; de Haan and Haagsma, 1988). Assuming this positive role, other scholars have looked to improve the regulations and standards of competition procedures (see, for example, Alexander et al., 1987; see also the Introduction to this volume).

More recently, the scholarship on architectural competitions has begun to diversify. Some have started to think about its role in design practices (Chupin, 2011; 2010), others have looked at the sociological networks competitions create and rely upon (Katsakou, 2011; Paisiou, 2011; Frey and Kolecek, 1995), still others have focused on forms of knowledge creation (Kreiner et al., 2011; Lipstadt, 2010; Silberberger et al., 2010; Nasar, 1999). It is now understood that the architectural competition is a component in a wider complexification of planning associated with new transnational geographies of city building and new private-public development partnerships (Van Wezemael, 2011; Van Wezemael et al., 2011; Malmberg, 2006). The competition format is not used for all and every component of city building, but it is especially prevalent with projects of a certain scale and profile. So routine is the competition that architectural historians have started to realize that they generate a significant number of not built schemes, themselves deserving of critical attention and archiving (Strebel et al., 2012; Chupin, 2008; Gomes Alves; 2008; Frey, 1998). As this existing scholarship suggests, and the chapter here aims to show, an architectural competition is more than simply a sieve of discernment that separates the wheat from the chaff, the good scheme from the not-so-good schemes. The ways competitions shape architecture are far more complex and furthermore they involve a range of geographical processes and outcomes. For example, awarding authorities routinely select construction sites and frame building briefs conceptually and programmatically. In return, many architectural offices now organize their work places to adequately resource the often fast-paced, high-pressure deadlines of competition opportunities. And as the arbiters of what does and does not get built, competitions articulate and stabilize mutually shared understandings of excellence in architecture. As Chupin (2011, p. 177) argues, competition jury boards actively shape and even design the projects they are assessing, evaluating and selecting during their meetings:

> [J]urors can be considered as the re-designers of the potential winning project, as if the judgement process required jurors to converge on a project, to the point where they can appropriate it, to make it theirs in a common decision. Since the winner is the 'product' of the judgement process, we can say that it is logically the 'project of the jury', as if the jury had designed it.

In what follows we want to focus on emergent geographies of architectural competitions. By this we mean, on the one hand, the spatial relations that competitions use to perform and, on the other hand, the spatial relations that competitive design procedures shape in process. We will do so by looking at jury board meeting from various angles, as they are the places were relevant spatial relations are tied together and maintained. Specifically, we address the procurement and administrative systems in which current competition practices are embedded in Switzerland, where we encounter the specificity of a highly regulated and standardized competition practice. This allows showing that understanding the geographical logic of the competition means going beyond regulations and norms, understanding them as practical in essence and looking at how they are made and sometimes unmade. Therefore, the chapter draws on data collected on the practices of twelve clients and competition organizers in Switzerland (Strebel et al., 2012; Silberberger and Strebel, 2011). In Switzerland, as in many other countries that are signatories to the World Trade Organization (WTO) Agreement on Government Procurement (GPA), every public awarding authority has to organize an international bidding as soon as estimated architects fees for a proposed scheme exceed an amount of CHF 350,000. As such the architectural competition is a forced part of the realization of publicly funded built schemes of a certain scale and value. In this context, our research aimed to understand competition regulations and instructions 'in action' (Suchman, 1987). The data was gathered in a study on the work practices of competition organizers, analysing internal documents, field notes from participatory observation and transcripts of interviews. We focused on a range of components in the competition process: how they are advertised, organized, monitored, participated in, publicized and sometimes aborted. This methodology enabled us to detail the local practices necessary to appropriately run and participate in competitions and, through this lens we have been able to better grasp how competition regulations and instructions are used and accomplished in the work situations studied (see for this approach: Lynch, 2002; 1993). In existing studies of architectural design and their regulatory contexts there is an emphasis on the 'normative' role of regulations, understood primarily through instructions given by professional associations and governmental institutions (Imrie and Street, 2011). Our approach brings into view a more complex and intimate relationship between design and regulation, at least in the context of competition systems.

The chapter highlights two selected geographical logics for how competitions and architecture are reciprocally and constitutively linked. Each reveals something about how instructions, rules and standards are used in competition procedures; how they are respected and adapted, as well as sometimes misused. The first geographical logic is how architectural competitions *structure their field of participants*, a part of which entails managing the interface between international market regulations and local labour markets. The second geographical logic relates to how the competition operates as a *place of knowledge production*. It is attentive to the spatial environment that jury boards use to take their decisions.

Structuring the field of participants: between international market regulations and local labour markets

One of the core arguments in support of the architectural competition is the way it opens a specific and necessarily localized building and design project to a global architectural community of expertise. Indeed, some would argue that it is only when a competition is globally open, that it can fully realize its potential as a device for enhancing the quality and innovation of architectural solutions proposed in response to a given brief. Interestingly, such claims made in purely design terms are not in conflict with the international regime for regulating public procurement processes. In contrast, as shown, the international competition is often at odds with local politics and regional markets. As noted earlier, in Switzerland any government procurement that exceeds a value of 350,000 CHF must be opened to international competition of tendering. That this is the case is an outcome of the Swiss government signing the GPA in 1996. As the WTO states: 'In most countries the government, and the agencies it controls, are together the biggest purchasers of goods of all kinds, ranging from basic commodities to high-technology equipment.' This market is, as the WTO notes, susceptible to 'political pressure to favour domestic suppliers over their foreign competitors'.[3] This statement captures the double rationalization of the GPA, which entered into force in January 1981. In the first instance the scale of public procurement is such that regulating this expenditure is a ready way the WTO can intervene in contexts where there is impropriety and inefficiency in the use of public funds, circumstances that in turn can impact negatively on confidence in government and good governance more generally. Second, as the WTO notes, 'public procurement of goods and services represents a major part of a country's market for foreign suppliers', so intervening in such procurement is also an instrument in expanding free trade into local markets through international competitive tendering.

As it happens, in the Swiss case, and as a brief look into the main architectural competitions carried out in 2011 reveals, only a small number actually displays a fully international field of participants. In fact, as we will show, actual competition practice falls short of the internationalizing goals of free trade frameworks enshrined by the regulation of public procurement under liberal market agreements. As we will see, this is not because of a wrong application of the rules, but due to local constraints set by the practical context of competition participation. In understanding this situation, Schmiedeknecht's (2010, p. 155) distinction between the 'routine' and 'exceptional' architectural competition is a helpful starting point. For Schmiedeknecht 'routine' competitions concern everyday or ordinary projects, which are not considered to be 'particularly glamorous', and where it is more important to 'fulfil functional requirements' than 'find spectacular formal solutions'. Housing schemes are a good example of routine competitions. In contrast, 'exceptional' competitions are 'perceived to be the place where the [architectural] avant-garde can show their credentials'. In Switzerland, we observe a different geography to routine and exceptional competitions. The former, which comprise the vast majority of competitions, have a larger proportion of national participants even though they are open for international bids as well. The latter often feature a truly international field of participating architectural offices.

This distinctive geography of participation is due to a number of reasons. First, a call for tenders is usually published in the official register of the city or the county or, as is the case in Switzerland, on a national online platform of public procurement (for example, www.simap.ch). Although the regulatory framework of the GPA requires an international reach in this call for tender stage the official advertisement of public competitions uses historically grown, and often parochial, advertising structures, which create difficulties for reaching potential competition participants abroad. These standard communication methods are thus very limiting, and it is common practice among organizers of exceptional architectural competitions (who want to enlarge and diversify their field of participants) to contact international firms and invest resources in a direct pitch of their competition opportunity. Organizers of routine competitions seldom take such initiatives. Nor would they, for example, bother to translate the competition brief into other languages. Second, participation in a competition is often linked to available architectural knowledge regarding local building law and the urban context. Participants also tend to base their decision to participate in a competition on 'sociological' knowledge about the client as well as the jury members. These fields of familiarity (with the competition clients and the local building and planning regimes) operate not only as a filter on who tenders for what kind of competition brief, but also as a shaper of the jury judgement. And they operate most potently with respect to 'routine' tenders. For certain kinds of international firm the resource demands of such required local knowledge are not outweighed by the budget or potential notoriety that might be gained through participation in a tender for a 'routine' competition. Furthermore, clients and organizers of routine competitions are prone to assume that local architecture offices better provide for local knowledge and organizational convenience (proximity, common language and building culture), and favour such firms. While routine competitions usually activate the local context, organizers of exceptional competitions – be they publicly or privately awarded – often actively and consciously generate and use an international field of participants in order to increase the visibility of their project and to 'sell' their design to the public (Sorkin, 2005; Collyer, 2004). So it happens that organizers of certain kinds of architectural competition can strategically evade an international field of participants despite the requirements of the GPA.

When speaking about how competitions structure a specific geographical field of competitors, it is also useful to understand that procurement regulation distinguishes between solution-orientated and performance-orientated types of tender. As the name implies the solution-based procurement applies to those situations where what is being sought and adjudicated upon is a solution to a problem or need. In architectural competitions the solution is presented in the form of a project. In solution-based tenders it is usual for the tendering company to anonymously submit their proposal: the assumption being that it is the solution, as opposed to the designer or provider of that solution, that is being adjudicated upon. In contrast, in performance-orientated forms of procurement it is usual for the candidates to be assessed on the basis of their track record and the commission to be awarded to the tenderer deemed to be most capable of providing the best solution to the task posed. One might assume that architecture is fully a solution-orientated procedure, and it is certainly true that in most contexts the tenders are submitted anonymously and judgement is of the scheme or project as opposed to the architectural firm. However, in the architectural competition, solution- and performance-orientated procurement types are intermingled. The client and organizer awards the commission on the basis of an assessment of the proposed architectural solution, but at the same time there is an understanding that a decision about an architectural solution will bring with it a business partner whose performance abilities will be central to how well or not the solution is realized. In practice there is no such a thing as a purely solution-orientated assessment. When jurors evaluate architectural schemes they do not and cannot judge them solely as solutions in isolation from the designing architect. This is clearly the case with 'exceptional' competitions that attract 'starchitects' whose design styles operate also as a company brand (McNeill, 2009). But this folding of solution and performance readily happens in the context of the 'routine' tender. For example when a participant in a conversation states: 'the corner of this house does not respond to the street' this is an adjudication of a scheme for a solution. But in an architectural competition the jury can also determine something of the potential performance from the presented solution, and they may, for example, be able to discern that 'the corner of this house is badly drawn'. In a jury board meeting both statements are used and jurors shuttle continuously between judging solution and judging potential performance. It might be thought that the anonymous solution-based tender system protects architecture from being susceptible to judgements that geographically localize the fields of participants. But the way various performance indicators (from a firm's track record through to its technical abilities) slip into and are expressed through the proposed solution means that no such guarantee can be assumed. In our research all juries adjudicating solutions drifted to judgements of performance. Furthermore, in the case of 'routine' competition tenders, this then allowed the jury to make assumptions and judgements about the ability of the 'anonymous' candidate to address local issues or mesh with local expectations about delivery. For our inquiry here into the geography of architectural competitions such slippage is significant. It shows that the geographical reach of the field of candidates cannot be adequately regulated for. Despite the intent of the WTO's international regulatory framework to keep 'local' interests at bay in large public procured architectural competitions, local building culture insistently shapes both the field of participants and final determinations.

Places of knowledge production: the spatial environment of decision making

Architectural competitions require judgements and the social organization that has the responsibility of adjudication is the jury. This is the second geographical logic of architectural production we wish to address, again thinking through the specifics of the architectural competition as it manifests in Switzerland. Every kind of decision making and planning activity, including those of the juries of architectural competitions, involve a continuous interaction with the local social and material context. We can learn about this powerful relational geography if we look closely at the places in which decisions are accomplished. Jury board meetings are usually pursued in unspecified and flexible places: conference rooms or big halls in which a number of plans and scale models, along with associated textual documents, are placed on show and examined by the jury, usually in the time frame of a single day. The venue for the competition adjudication is chosen rather pragmatically, often on the basis of the number of projects to be displayed, and taking into account practical matters such as the location of the venue with respect to public transport. Presence on the building site is not primarily an issue. The jury venue is a temporary workplace for a group of experts, representatives of the client and others. Furthermore, there are often stages in this jury process, pre-qualification screenings as well as final jury meetings.

In what follows we wish to focus on one particular instance of such a workplace and look at its spatial arrangements and how they contribute to the architectural adjudication. Looking closely at Figure 1.1, which shows an instance of a pre-qualification jury board meeting, we can see that the space is arranged such that there is a 'room in a room': a rectangular space of approximately 16 by 20 metres was arranged inside the 2236m2 hall using movable walls. Such organizational aspects although seemingly minor, nonetheless enshrine something of the values and rules that produce the architectural competition. One aspect of the jury room is that it is arranged so as to give each entry the same amount of space and so an equal position in the jury room. The work inside the jury space is the work of a designated group of experts and representatives of the client, who are officially and publicly entrusted with the responsibility of coming to an accurate and robust (because unchallengeable in principle) decision. Pre-qualification meetings, such as the one depicted in Figure 1.1, are performance-orientated: the aim is to produce a final and decisive shortlist of architectural firms to be invited to hand in schemes in the main part of the competition. The selected architects or teams then will be invited to develop and submit their projects for the competition, which is solution-orientated. All that is delivered to the public at this stage is an alphabetical shortlist of architectural offices, which hides all the arguments and discussions of the jury board sessions. In the pre-qualification assessment stage the jury has no obligation to report to the public its shortlist outcomes (why this architectural office and not another) or even the criteria of adjudication. This exclusive arrangement can be thought of as a powerful translation process (Callon, 1986), through which a list of potential participants and, importantly, architectural offices that the jury thinks possess the competences to execute the project should they win, is

Two geographical logics 37

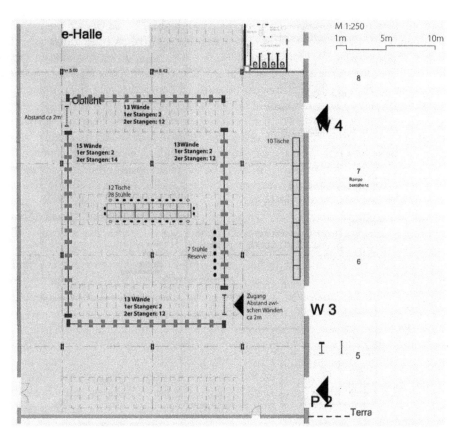

Figure 1.1 Extract from a selected competition preparation document. The furnishing of the space in which the preselection meeting happened is planned in every detail. The layout of the space is presented in a drawing (original scale 1:250) and elements of the set up listed as follows: 54 movable pinboards, 8 corner poles, 50 poles, 22 tables, 28 chairs, 7 back-up chairs.
Source: Courtesy of the Department of Planning and Architecture, City of Basel.

materially stabilized and displayed in a way that the jury board can collectively and legitimately speak about, as is done in the final jury reports.

Let us look even more closely at the work that happens inside the quadrangle-shaped place of adjudication. Another feature of that work is the mobility of the jury board. The jurors always move to schemes as a group, engaging collectively with each of the models and plans. They almost always stand as they gather round the models and plans that represent the scheme or, as in this case, round the documents that present the firm and its portofolio. One of us, who was given the opportunity to participate in the pre-selection meeting, observed the jurors first circuiting the schemes in groups of three, assessing, tagging and labelling the entries using a predefined evaluation system. In a second phase the entire jury went from scheme to scheme and assessed collectively as a jury board each entry, again according to a predefined evaluation system. Towards the end of the pre-evaluation session members sat around the table in the middle of the room and further discussed the entries, comparing them and progressively reducing the number of candidates. At this final point of discussion, much of the order of the earlier evaluative stages gives way to a messier process in which arguments and counter arguments go in various directions among the jury, although this contentious action is still aimed at producing a common result. We can see here one of the specific features of the modern competition. Any arguments from members within the jury itself should not be traceable to the individual once the final jury decision has been made, for the decision must appear as one that is made collectively. The production of a single collective decision is achieved by the jury sessions happening in spaces not open to the public. The reports written in this space about the determination do not indicate anything of the messy contention but deliver to the public realm a stable, collectively agreed argument. Another dimension of the organization of the pre-selection process supports the production of collectively attained determinations. The jury members are not given information about the entries before the pre-selection meeting. This ensures that the assessment process is developed in a cooperative arrangement among the jurors themselves, and reduces the risk of jurors coming into the pre-selection with hardened opinions, possibly formed in collaboration with others outside the jury itself. Here we can see clearly that the ephemeral space, specifically set up for this one jury board meeting, supports the competition processes, its methods of collective decision making and its rules for anonymity.

That the adjudication process of contemporary architectural competitions happens behind closed doors is interestingly inconsistent with the history of the modern architectural competition. As Bergdoll (1989) shows in his historical account, the democratic spirit seeded by the French Revolution transformed the traditionally closed architectural competition of the Ancien Régime into open, and so publically accountable, events. Democracy forced upon the architectural competition a new public function. In the aftermath of the French Revolution the geographical articulation of competitions was such that participants, the items under judgement, and the act of adjudication were all assembled and performed in one public space. We do not want to generalize this historical development; however, we can see hints of the emergence of an open and democratic spatial and social arrangement in a late nineteenth century drawing of the Salle Melpomème by Beaux-Arts student Alexis Lemaistre (1889) (Figure 1.2).

The image depicts a large space that was designed especially for competition entries to be hung and judged in what were monthly competitions. Entrants registered their drawings at a table in the middle of the room, before their works were hung and then judged. In this instance the place of the competition, and the routines of organization it supported, were not only pluralistic in expression, but also in judgement (Bergdoll, 1989). This is a single example, however – as Bergdoll (1989) further develops – it precedes changes in public procurement that happened after, defining the open competition to be the main tool for all municipal commissions of buildings, engineering and art works. Those competitions had to be publicly announced and anyone could submit a proposal. Bergdoll highlights other differences to the competition as we encounter it, at least in the western world, today. Back then competitors were supposed to elect the jury board themselves and the judgements of the board were openly debated in public after it had taken its decision.

It is clear that the modern intent of the architectural competition is to instate routines that support fair and independent judgement in the name of a more publicly

Two geographical logics 39

Figure 1.2 Registering of competition entries in the Salle Melpomène, Ecole des Beaux-Arts in Paris.
Source: Image extract reprinted from Alexis Lemaistre, 1889 (the entire image is reprinted in Lipstadt, 1989, p. 10).

accountable outcome. But ironically the contemporary competition, at least as it is manifest in the Swiss examples that we have studied, has less public access and input than the democratic competition after the French Revolution as described by the historian. There are few opportunities for public participation or the expression of a diverse range of views in the decision making process. The spatial setting of the contemporary jury board meeting today produces a strictly restricted and controlled public involvement. This is not to produce a political space such as the Salle Melpoème under such conditions that the power to decide is given to a restricted and exclusive number of people. The spatial and social organization of the jury board meeting supports a small group of state-designated experts withdrawing from the public in order to carry out their assessment and decisions based on their specialist knowledge. The expectation is that the jury room functions as a space of expertise in which authoritative and trusted knowledge is produced. But does it? Can it? By way of comparison we might think of the scientific laboratory, which Shapin (1998) referred to as a 'place of knowledge' precisely because it produced trusted knowledge able to influence and convince others. The authority of the scientific laboratory as a place of knowledge was the outcome of a sustained association between a socially removed material space (the laboratory) and a group of experts (the scientists who routinely used the laboratory). The ability of the laboratory to be a respected 'place of knowledge' is produced by that sustained association and the accretion of proofs arising from the experiments held within. Furthermore, the laboratory and the experiment designs enshrined in it, are a key to the replicability of the experiments and so the veracity of the knowledge produced. The contemporary architectural jury room may have aspirations to such authority, but it functions in a different way. As in the laboratory, the experts in the jury room are expected to create knowledge (their judgement) that is authoritative and can keep that authority in place as it circulates away from where it was made to its varied destinations. But unlike the laboratory, the contemporary architectural jury room is an ephemeral, temporary space in which the experts are gathered for only a short period of time. Furthermore, after the jury session, this 'place of knowledge' dissolves: the experts go back to their various posts, the material arrangement of the place is taken apart and the entries are packed away. As such, there is no possibility given by this space for others to collaboratively assess, compare and evaluate the projects. Replication of the reasoning inside the decision making process is simply not possible, for it dematerializes. This means that in the case of the architectural competition if there is a public challenge to the decision, or doubt, there is no recourse to the originating 'place of knowledge'. The jury board places are assembled to support and facilitate the expert jury making their aesthetic judgement and to enable them to display that decision authoritatively, be that a list that describes the jury and their expertise or the jury report on the decision. But these translations of the act of adjudication are always susceptible in part because no member of the jury board has the possibility to rebuild (replicate) the decision making process and so to verify the contested decision. We can see here very clearly how aesthetic judgement differs from epistemic judgement.

More recently the ephemeral and closed jury space has had the potential to become more 'permanent' and open. Non-public clients who organize competitions, are starting to use digital tools as part of the competition procedure. Digital tools simplify the process of assembling a competition by allowing remote entry submission and easier sharing of information among jury board members. Most competition organizers see such tools as cost-saving because they help to slim down competition procedures. In contrast, architects and expert members of jury boards take a rather more conservative position in relation to the digitalization of competition procedures. Architects, for example, fear they will loose control over the ways their projects are displayed to the jury board, especially if their entry is submitted only through digital assets, rather than also a built model, which remains a compelling part of the architect's art of persuasion. Similarly, jury board members are concerned that digitized dissemination would break apart the spatially emplaced and materialized cooperative work enshrined in the jury session in which all jury members view all projects on one occasion in one room. In part because of this resistance, the current development of digital tools in architectural competitions is largely aimed at supporting the administration of the submission process as opposed to the jury process itself (Strebel et al., 2012). For example, digital toolboxes might support competition organizers in coordinating and centralizing the administrative workload that goes with the organization of an architectural competition. Early experiences with such tools show that digital support does not enhance the openness of the competition process, but it does distribute the competencies surrounding the jury board meeting differently; for example, it redistributes the question of by whom and when in the process the digital images are printed. Although digitization is transforming procedural aspects of the architectural competition, at this point it does not seem set to solve some of the contradictions raised above about the closed and unreplicable nature of the judging process and the impact this has on the authority of the determination.

Let us briefly go back to the pre-qualification, a performance-orientated procedure, which was described above. Usually, the applicant has to fill in a pre-formatted form (in pdf or word format), and has to print the form and send it together with two to three reference projects printed on an A3-sheet in the mail to the competition organizer. The competition organizer will unpack the application portfolio and use the A3-prints in the jury sessions. As new web-based tools today allow submission of portfolio and reference projects digitally, competition organizers receive applications in pdf-format, which eventually they could send to jury board members for preparation or, what most of the organizers would do, print them and proceed to arrange the prints in the old fashioned way in the closed jury space. The digitalization of the jury board process here does not provide for a new spatial arrangement, but for a different distribution of competences. Printing on paper is now the task of the competition organiser and the participant loses control over the materiality of his or her entry, which in architectural terms can be very problematic.

Conclusion

We stated at the beginning of this chapter that within the professional field of architecture and urban design the architectural competition enjoys the highest confidence and is praised, very often without further scrutiny, as the ideal condition of a design procedure. Using Bruno Latour's terms we could state that the competition within the profession is understood and promoted as an intermediary, which is something that 'transports meaning or force without transformation: defining its inputs is enough to define its outputs' (Latour, 2005, p. 39). As we scrutinize this uncritical appreciation and sometimes celebration in this chapter we have been looking at competitions not as intermediaries but as what Latour calls mediators, whose 'input is never a good predictor of their output; their specificity has to be taken into account every time' (p. 39). We claim that the real quality and force of the competition becomes visible when the regulations, procedures, tools, people and plans used in competitions are treated as such mediators. This might look like splitting hairs and another attempt to force and promote an approach in social theory, however, we claim that with the understanding of the competition procedure as a mediator that 'transforms, translates, distorts and modifies' (p. 39) architectural and urban projects, we gain a better understanding of how the competition – as one specific and selected step in the construction process – is embedded and therefore effectively relates and shapes building and architecture end-to-end. One obvious example of the ways architectural competitions shape the relational field post-competition is the fact that in an anonymous solution-orientated competition awarding authorities will never know the names of participants and competing architects during board meetings. The conversation between selected architect and client starts therefore very late in the process, only when the project is defined. This can develop as a severe problem in upcoming negotiations (Nasar, 1999), which in many cases, depending on the size and advancement of project detailing and construction of the building, can last for months and years, and raise concern if it does not work from the beginning.

Concerning the first point of the chapter, it is important to understand how even competitions that are declared to be 'open' and which explicitly do not restrict participation in terms of, for example, office size, track record or geography, structure the field of potential participants, and this despite WTO procurement regulations. Larger architectural firms from countries outside Switzerland have recognized such hidden selection processes, have taken this into account and have consequently developed new strategies to participate in competitions in Switzerland. The problem here is that such firms can set up a Swiss branch of their company and hire locally trained architects to participate in competitions. This significantly increases their chances of winning, as the local branch allows them to perform in the Swiss context, as local knowledge on, for example, building regulations can be activated. Concerns surrounding this dynamic were recently raised in a focus group workshop by the leader of an architectural office that has been successful in various competitions in Switzerland. The problem for this architect is not just the increasing competition that such tactics by international firms provide, but even more importantly the possibility that the Swiss branch will be closed after the firm has won the competition. The winner will work abroad, not least because the salaries they pay will be much cheaper, meaning the awarding authority will have to communicate with the responsible architect at distance and very likely the client will still have to work with people who do not know the local construction market.

In the second part of the chapter, we have discussed the ephemeral character and non-traceability of decisions made in architectural competitions. Our analysis of the ways jury board meetings work as places of knowledge has clearly revealed that the practical and geographical organization of the competition procedure is crucial to the stability and consistency of the argumentation that leads to the winning project. Practical aspects such as the erection of pin boards for plans, the arrangement of tables and chairs shape the outcome of the jury board meeting and support collective decisions taken and the inter-subjective agreement on the winning project. In this regard and going once more beyond the jury board meeting, we can show that good competition organization is not to be understood as an coherent application of regulations and standards as defined through the public procurement system, but involves practical aspects of how to unfold competitions in the particular local context of the project to be developed and built.

Notes

1 www.sia.ch/de/dienstleistungen/artikelbeitraege/detail/article/der-architekturwettbewerb/ (last retrieved on 9 December 2016).
2 Wettbewerbsstandard Architektur – WSA 2010.
3 www.wto.org/english/thewto_e/whatis_e/tif_e/agrm10_e.htm (last retrieved on 5 February 2013).

References

Alexander, E. R., Witzling, L. P., and Casper, D. J. (1987). Planning and urban design competitions: organization, implementation and impacts. *Journal of Architectural and Planning Research*, 4, pp. 31–46.

Bergdoll, B. (1989). Competing in the Academy and the Marketplace: European Architecture Competitions. In H. Lipstadt (Ed.), *The Experimental Tradition*, New York: Princeton Architectural Press.

Callon, M. (1986). Some Elements of a Sociology of Translation: Domestication of the Scallops and the Fishermen of Saint Brieuc Bay. In J. Law (Ed.), *Power, Action and Belief: A New*

Sociology of Knowledge? Sociological Review Monograph 32. London: Routledge and Kegan Paul.

Chupin, J.-P. (2008). Documenting Competitions, Contributing to Research, Archiving Events. In D. Peyceré, F. Wierre and C. Koch (Eds), *Architecture and Digital Archives. Architecture in the Digital Age: A Question of Memory*. Gollion: Infolio.

Chupin, J.-P. (2010). *Analogie et théorie en architecture: De la vie, de la ville et de la conception, même*. Gollion: Infolio.

Chupin, J.-P. (2011). Judgement by design: towards a model for studying and improving the competition process in architecture and urban design. *Scandinavian Journal of Management*, 27, pp. 173–184.

Collyer, S.G. (2004). *Competitions: Competing Globally in Architecture Competitions*. Chichester: Wiley-Academy.

de Haan, H., and Haagsma, I. (Eds) (1988). *Architects in Competition: International Architectural Competitions of the Last Two Hundred Years*. New York: Thames and Hudson.

de Jong, C., and Mattie, E. (1994a). *Architectural Competitions 1792–1949*. Cologne: Benedikt Taschen.

de Jong, C., and Mattie E. (1994b). *Architectural Competitions 1950–today*. Cologne: Benedikt Taschen.

Frey, P. (1998). *Histoire et archives architecturales: éléments méthodologiques et informatiques: le fonds Alphonse Laverrière aux Archives de la Construction Moderne* (PhD Thesis). Lausanne: École Polytechnique Fédérale.

Frey, P., and Kolecek, I. (Eds) (1995). *Concours d'architecture et d'urbanisme en Suisse romande. Histoire et actualité*. Lausanne: Editions Payot.

Gomes Alves, L. (2008). *L'archivage numérique des projets 'Europan' comme situation d'analyse scientifique du concours d'idées en architecture (Mémoire de maîtrise)*. Montreal: Laboratoire d'étude de l'architecture potentielle.

Imrie, R., and Street, E. (2011) *Architectural Design and Regulation*. Oxford: Wiley & Blackwell.

Katsakou, A. (2011). *Recent Architectural Competitions for Collective Housing in Switzerland: Impact of this Framework on Architectural Conception and Innovation* (Doctoral Thesis). Lausanne: École Polytechnique Fédérale.

Kreiner, K. (2010). Architectural Competitions: Empirical Observations and Strategic Implications for Architectural Firms. In M. Rönn, R. Kazemian and J. E. Andersson (Eds), *The Architectural Competition: Research Inquiries and Experiences*. Stockholm: Axl Books.

Kreiner, K., Jacobsen, P. H., and Jensen, D. T. (2011). Dialogues and the problems of knowing: reinventing the architectural competition. *Scandinavian Journal of Management*, 27, pp. 160–166.

Latour, B. (2005). *Reassembling the Social: An Introduction to Actor-Network-Theory*. Oxford: Oxford University Press.

Lemaistre, A. (1889). *L'École des Beaux-Arts desinée et racontée par un élève*. Paris: Firmin-Didot & Cie.

Lipstadt, H. (Ed.) (1989). *The Experimental Tradition*. New York: Princeton Architectural Press.

Lipstadt, H. (2010). Experimenting with the Experimental Tradition, 1989–2009: On Competitions and Architecture Research. In M. Rönn, R. Kazemian and J. E. Andersson (Eds), *The Architectural Competition: Research Inquiries and Experiences*. Stockholm: Axel Books.

Lynch, M. (1993). *Scientific Practice and Ordinary Action: Ethnomethodology and Social Studies of Science*. New York: Cambridge University Press.

Lynch, M. (2002). The Living Text: Written Instructions and Situated Actions in Telephone Surveys. In D. W. Maynard, H. Houtkoop-Steenstra, N. C. Schaeffer and J. van der Zouwen (Eds), *Standardization and Tacit Knowledge: Interaction and Practice in the Survey Interview*. New York: John Wiley & Sons.

Malmberg, C. (Ed.) (2006). *The Politics of Design: Competitions for Public Projects*. Princeton: The Trustees of Princeton University.

McNeill, D. (2009). *The Global Architect: Firms, Fame and Urban Form*. New York: Routledge – Taylor & Francis.

Nasar, J. L. (1999). *Design by Competition: Making Design Competition Work*. Cambridge: Cambridge University Press.

Paisiou, S. (2011). Four performances for the new Acropolis museum: when the politics of space enter the becoming of place. *Geographica Helvetica*, 66, pp. 33–41.

Schmiedeknecht, T. (2010). Routine and Exceptional Competition Practice in Germany as Published in *Wettbewerbe Aktuell*. In M. Rönn, R. Kazemian and J. E. Andersson (Eds), *The Architectural Competition: Research Inquiries and Experiences*, Stockholm: Axl Books.

Shapin, S. (1998). Placing the view from nowhere: historical and sociological problems in the location of science. *Transactions of the Institute of British Geographers*, 23, pp. 5–12.

Silberberger, J. (2012). Jury Sessions as non-trivial machines: a procedural analysis. *Journal of Design Research*, 10, pp. 258–268.

Silberberger, J., and Strebel, I. (2011). *Zwischenbericht zu den Fallstudien 'Wettbwerbsorganisation' – KTI-Projekt 11834.1 PFES-ES Wissenssystem Wettbewerb*. Zurich: ETH Wohnforum – ETH CASE.

Silberberger, J., van Wezemael, J. E., Pasiou, S., and Strebel, I. (2010). Spaces of knowledge creation: tracing 'knowing in action' in jury-based decision-making processes in Switzerland. *International Journal of Knowledge-Based Development*, 1, pp. 287–302.

Sorkin, M. (2005). Democracy Degree Zero. In Österreiche Gesellschaft für Architektur (Ed.), *UmBau 22: Wettbewerb! Competition!* Vienna: Technische Universität.

Spreiregen, P. (1979). *Design Competitions*. New York: McGraw Hill.

Strebel, I., Silberberger, J., and Raschpichler, D. (2012). *IT-Konzept zum KTI-Projekt 11834.1 PFES-ES Wissenssystem Wettbewerb*. Zurich: ETH Wohnforum – ETH CASE.

Suchman, L. (1987). *Plans and Situated Actions: The Problem of Human–Machine Communication*. Cambridge: Cambridge University Press.

Van Wezemael, J. (2011). Forms, places and processes: tracing geographies of architecture through design competitions. *Geographica Helvetica*, 1, pp. 2–4.

Van Wezemael, J., Silberberger, J. and Paisiou, S. (2011). Assessing 'quality': the unfolding of the 'good' – collective decision making in juries of urban design competitions. *Scandinavian Journal of Management*, 27, pp. 167–172.

Volker, L. (2011). *Deciding about Design Quality: Value Judgements and Decision Making in the Selection of Architects by Public Clients under European Tendering Regulations*. Leiden: Sidestone Press.

2 The competition between creativity and legitimacy

Kristian Kreiner

Introduction

Competition plays an almost sacred role in modern society. It is believed to be a social mechanism (Hedström and Swedberg, 1998) that ensures efficiency and fairness in the allocation of scarce societal resources and therefore to afford both economic growth and rationality. For instrumental and moral reasons, the best man (or woman) should win and get ahead, competition offering such a promise. In practice, social mechanisms have to be organized, staged, framed and managed, the ways in which this is done impacting on their real, observable effects. When the social mechanism we call competition is turned into an observable event, that is *a* competition, the *intended* effects become tasks for the organizers, contestants and audience to achieve. However, in spite of all effort, the resulting effects may not prove to be identical with the intended ones. Social mechanisms may guide and inspire the ways in which competitions are conducted, but whether efficiency and fairness ensue is fundamentally an empirical question. This question will be examined in the context of an architectural competition in Denmark.

Architectural competitions represent a highly institutionalized field in which the social mechanism of competition becomes interpreted, adapted and instantiated. The general concerns for efficiency and fairness are translated into the aim of letting the architects compete for the best ideas and solutions in relation to some building project and letting the best one win the contract for detailing the design and overseeing the subsequent construction. Efficiency includes a concern for creativity, winning implying having the most original conceptualization of the design task and a feasible solution to it.

The design and management of architectural competitions reflects a certain special feature of this competitive field. The architectural competition is an early step in an often large, long-term investment project. At the time of the competition, the object does not exist in material form but is made imaginable and current, first, in the form of a competition brief that defines the general requirements and parameters for the architects' work, and, second, in the form of the architects' specific design proposals. The translation of the future material building into a current abstract design is necessarily partial, creative and professionally demanding. Such translations are what architects have learned to do and are paid to do in order for the client to make informed investment decisions. Translation is a creative but also uncertain technology, which is why architectural designs can never simply be deduced from the needs and wishes of the client. To ask several architects to produce a design is bound (and meant) to produce highly different translations. When architects' efforts are coordinated in time and space, the client can be presented with several alternative translations and choose the best of the alternative designs for later realization.

Conceivably, the mere fact that the practicalities of organizing architectural competitions have become highly institutionalized indicates a history of difficulties with accomplishing the *intended* fairness effects of competitions. For example, clients may be unable to recognize and rank the translations of their needs and wishes in the various design proposals and may feel tempted to use the reputation of the architect as a cognitive decision heuristic (Kahneman, 2011). In practice, the client may have stronger preferences for (or against) specific architects than for their respective design proposals. The institutionalized role of a competition jury may serve as illustration of the ways in which the risk of non-rational choices and unfair outcomes are mitigated. For example, in most cases, the client representatives on the competition jury are supplemented with architects appointed by the Danish Association of Architects to ensure a professional and unbiased assessment of the proposals. Furthermore, the identity of the architects behind the design proposals is normally kept secret until the jury reaches a decision. Professional experts on the jury and the anonymity of the design proposals make it harder to use illegitimate concerns and criteria in appointing a competition winner.

One additional way of ensuring fair architectural competitions is to be precise about the design task. At least in Denmark, it is often claimed that the competition briefs are ambiguous and that efforts are wasted (and the chance to win ruined) when architects misunderstand the task. To combat ambiguity and prevent misunderstanding, competition briefs have become increasingly voluminous and detailed. Delimiting the design task and narrowing the solution space would seem to reduce the opportunity for creativity. It has been argued, however, that more words and specifications have failed to alleviate ambiguity. On the contrary, another layer of ambiguity is added when the necessary prioritization among a multiplicity of demands and requirements entails subjective judgement (Kreiner, 2012).

This being the case, it may be surprising that the outcomes of architectural competitions are seldom contested, at least legally. Competitions produce few winners and many losers, the latter perhaps easily left feeling misled by the ambiguity of the competition brief. However, while outcomes may be debatable, it is generally accepted that the winner is the right one. One empirical study found that the losers did not feel misled but acknowledged having misread the brief (Kreiner, 2013). By assuming responsibility for not having read the brief properly, the outcome transpires as correct and fair.

However, occasionally competitions produce not only winners but also controversy, as the case under analysis demonstrates. There is little reason to doubt that the competition was professionally organized. Without escaping the inherent ambiguity of competition briefs, it had all the qualities and elements to encourage the belief that the competition was based on a well-defined, shared design task. The stage was set and the rules were in place for an ordinary, that is harmonious, competition, with highly professional architects and jury members participating.

Why, we must ask, did such a well-organized competition produce so much controversy? This question implies another question, namely what did the organizers do to avoid or subdue the controversy? Our ultimate question follows from this: is it possible that the difference between this competition and ordinary competitions is not the controversy in itself but the ineffectiveness with which these controversies were managed in the present case? If this is the case, we need to theorize the managerial mechanisms by which inherent controversies are made to appear harmonious in ordinary cases.

Before addressing these questions, we will account for the methodology of the study and describe the case in some detail.

Methodology

This study involves an architectural competition for a new building for the Maritime Museum of Denmark. The museum is run privately but because it receives public subsidies, it is governed as a public institution. Professionals planned and managed the competition, but in the name of the museum and subject to the rules and regulations of public procurement. We refer to the museum in relation to the competition as 'client'.

The study builds on desk research and interviews. We collected data on the historical facts of the competition, its prior circumstances, the actions of various participants and bystanders, and the succession of events that constituted the competition and its aftermath.

Readily retrievable written documentation simplified the desk research as legal proceedings resulting from the competition documented many facts in the case. Due to the controversy, the press covered the story and kept reporting on the project long after the competition. The fact that the various parties were in agreement about what actually happened also made the desk research more straightforward. Even during the court proceedings, the client acknowledged the correctness of the opponent's account of what happened. To an unusual extent, all parties were in agreement about what had happened, who had done what, and with what consequences.

On top of this desk research, we conducted extended interviews with various individuals directly or indirectly involved in the competition. All interviews were transcribed and then approved by the interviewees. The interviews explored in greater detail the historical events, the intentions and strategies of the actors, and also some general conditions and relationships in society and the building sector. We interviewed representatives from funders, the trust fund, the museum board and the management of the Danish Association of Architectural Firms taking the client to court, as well as the consulting engineer organizing the competition. A few actors on the periphery of the competition were interviewed for background information. Some of the interviewees changed roles during the competition.

In our view, case studies lose value when confined to disciplinary dry docks, thus calling for an eclectic approach. Our choice of theoretical perspective was premised on the idea that the concrete case and historical course of events should highlight general conditions and circumstances in architectural competitions and in society at large. Theories add insight to visible, that is, empirical, facts. For instance, theoretical reflection is required to recognize the significance of the distinction between competitions with private and public clients when studying a specific case. We also used theoretical reflection to produce a realistic but counterfactual history of the documented competition to add new perspectives and significance to the actual and historical events.

The case: an architectural competition for the new Maritime Museum of Denmark

In October 2013 the Queen of Denmark opened the new Maritime Museum of Denmark, which received rave reviews everywhere and was even mentioned on the *New York Times*' prestigious list of 52 Places to Go in 2014. However, there was a fly in the ointment: the Danish Association of Architectural Firms took the client to the Complaints Board for Public Procurement, which unanimously ruled that the winner of the architectural competition leading to this astonishing amount of recognition was not legitimate. The ruling did not surprise the client, who admitted to having broken the rules. And, despite the ruling, nothing hindered the successful implementation of the winning design proposal.

It seems that something is still rotten in the state of Denmark, as Shakespeare's Hamlet famously once declared, though the case appears to be an exception to the rule. Rarely are clients taken to court for mishandling architectural competitions. Local competitions of this nature also seldom produce architectural designs that receive global recognition and acclaim. Most of all, it is unusual that the concern for architectural quality and the concern for legitimacy collide so violently. A general assumption is that a properly designed and executed competition will ensure that the best design proposal wins. As a rule, the two concerns are perceived as going hand in hand, but our study reminds us that this is not necessarily the case. Had the concern for procedural legitimacy prevailed, an inferior design proposal would have won, but with almost no chance of being built. Since the obviously best proposal was appointed winner, the concern for architectural quality prevailed, but the trust in the fairness of architectural competitions was shaken by the events, at least temporarily.

Below, we will give an account of the competition process. Special emphasis will be given to the ways in which some actors tried to avoid a confrontation between the concern for architectural quality and the concern for the legitimacy (and fairness) of the competition. Emphasis will also be given to how concern for architectural quality was granted primacy even after the court had disqualified the competition and the appointed winner.

While hopefully interesting of its own accord, the detailed account of this outlier of a competition is aimed at enabling us to argue that the concern for quality and the concern for legitimacy are more generally at odds, and that in 'ordinary' competitions, the two concerns are successfully rationalized to work in unison in the consummation of architectural competitions.

The client and the historical background for the competition

Nothing in the history of the Maritime Museum of Denmark indicated that it would be included on the *New York Times* 52 Places to Go in 2014, their road to fame being rather indirect and, in many ways, faint. For years, this private museum enjoyed a quiet existence within the famous walls of Kronborg Castle, also known as Hamlet's Castle, in Elsinore, Denmark. In 2000, the castle became a UNESCO World Heritage Site, prompting a major restoration requiring relocation of the museum – which did not fit into the Danish Ministry of Culture's vision for the future of the castle, and was asked to find a permanent location outside its walls.

Creativity versus legitimacy 49

The sudden and unexpected exposure to public attention spelled difficult times for the museum, the Ministry of Culture reconsidering its public subsidies and suggesting a merger with another small museum in western Denmark. In the midst of this identity crisis, strong interests came to the rescue of the museum. Denmark has a long history as a maritime power; its shipping industry, headed by the global player Maersk, plays an important role in the world economy. This position translates into a strong political role in Danish society further enhanced by the powerful philanthropy of highly wealthy private foundations controlled by the shipping industry. As natural allies of the museum, shipping companies and their philanthropic foundations soon changed its gloomy prospects for the better.

Next to Hamlet's Castle, an abandoned historic shipyard in the harbour of Elsinore was turned into a civic and cultural centre. Another private Danish foundation spent a considerable sum of money on clearing and restoring the waterfront to create a strong visual link between the castle and the new centre. The local authorities suggested that the museum could use a dry dock left idle after the project as an, admittedly challenging new home. This location right between the centre and the castle raised strong concerns, creating the stipulation that the museum had to be built completely underground, quite literally within the perimeter of the dry dock. This stipulation would soon become fateful in an unforeseen manner.

The competition design

In January 2007, the museum ran an invited architectural competition. Since the museum receives part of its operating budget from the Ministry of Culture, it is subject to rules stipulating that commissions cannot be awarded without a compulsory open competition. With substantial estimated costs for the project (DKK 130 m, EUR 17.5 m), the competition had to be announced to all eligible architects in the EU.

Thirty-eight architects responded to the invitation to apply to be a participant, five of whom were selected based on a variety of criteria asserted to create, for example, a balanced representation of young versus established and domestic versus European architects, though the specific criteria were not defined.

The competition was a single-stage, sealed-bid invited design competition specifying the task of fitting a 4,500 m2 museum into the dry dock. The five contestants were guaranteed a fee for their participation, provided that their submitted design proposals were in compliance with the brief. This type of remuneration generally covers some but seldom all of the architect's expenditures. The winner would receive a cash prize, but, more importantly, the prospect of most likely being granted the design contract. The runner-up proposal would also receive a small cash prize.

A highly esteemed engineering firm was hired to assist the client in organizing and managing the architectural competition. Some of the above-mentioned philanthropic foundations secured funding. Informally, funding the architectural competition also indicated a willingness to finance construction of the winning proposal. To align interests, the funding foundations and the foundation concerned about the visual link between the castle and the civic and cultural centre were represented on the competition jury. The Danish Association of Architects appointed further members. Naturally, the client was also represented on the jury.

The brief comprised extensive requirements, as the dry dock was a challenging site architecturally and from an engineering perspective. The narrowness of the dock put

50 *Kristian Kreiner*

severe limitations on the museum's design, the necessary technical solutions putting severe constraints on the budget. The criteria in the brief the jury was to follow to appoint a winner involved:

- The overall architectural, aesthetic, functional and technical quality of the design
- The estimated costs of the construction within the limitations of the budget of DKK 130 m (EUR 17.5 m)
- The quality and qualifications of the professional staff that would be responsible for detailing the design and overseeing the construction
- The operational and maintenance costs of the facilities

The architects had less than six months to prepare design proposals. A few opportunities to ask for clarifications were made available initially and resulted in myriad questions, indicating the difficulty of the task. The participants pointed out that specifying future professional staff would breach the anonymity stipulation. Consequently, the client decided to delete this criterion. When asked how closely the perimeter of the dock should be respected, the client answered unequivocally that it should be taken literally, both vertically and horizontally. The competition was launched on this premise.

Controversies and recoveries

When the five design proposals were received, chaos seemed to ensue, confining the museum to the perimeter of the dock apparently necessitating multiple architectural, aesthetic and functional compromises. According to our interviewees, the jury showed little appreciation for any of the design proposals, with the exception of one, which proposed building the museum into the sidewalls of the dock, leaving the dock as an open space. Such a design would mean costly engineering problems but would require fewer architectural compromises, garnering it an extremely high score on the first (and probably most important) evaluation criterion but causing it to completely miss the mark on the second one. Obviously, this design was not in compliance with the oft-repeated requirement that the perimeter of the current dry dock was not to be broken.

The dilemma for the client and the jury was clear: the legitimate design proposals were not attractive and the attractive proposal was not legitimate. Implementing an obviously inferior design solely because it complied with the brief seemed impossible. Similarly, going ahead with an attractive design that so blatantly ignored the rules also seemed unfeasible. Finally, cancelling the competition and beginning anew, especially due to time constraints, also seemed nonsensical. Without a new venue when evicted from the castle, the museum would most likely be sent into the arms of an unappealing provincial museum. Moreover, the money used to organize the competition would have been wasted, and conditional grants would not have a project to fund.

This apparent stalemate ended when, out of a sense of necessity, the client[1] was forced to act inventively. The first attempt at a recovery involved the client sending a letter to all five architects asserting that none of the design proposals complied with the second criterion (costs) in the brief, which was why none of them would be entitled to receive their participation fee. The client, however, did offer to take this criterion out of the equation to allow them to receive their fee on the condition that they would forego the right to contest changes in the competition and its ultimate outcome.

De facto, the client's suggestion would neutralize all evaluation criteria, leaving the field completely open to pick the desired winner.

Importantly, at this point in the process, information was distributed unequally. The jury and the client were only familiar with the details of each proposal but not the identity of the architects. The architects thus had to respond to the client's suggestion based merely on their own proposal. All but one architect opted for guaranteed compensation and agreed to the client's proposal. The remaining architect refused the offer, seeing the cost disqualification as unfair, due to the conviction that that architect's design proposal stayed within the stipulated budget. Estimating costs in the early stages of a project can be difficult; disqualifying all entries on this account is thus surprising and at any rate unconventional. Only one issue was unequivocal, the cost of building the museum into the walls of the dry dock was in a completely different ballpark than what the brief prescribed.

The client's second recovery attempt was more forthright than its first. The client simply stated that since the design proposals blatantly neglected the budget criterion, the jury was entitled to do likewise. The jury then publically announced that the winning design proposal was the one that failed to stay within the perimeter of the dry dock. Two second prizes were awarded and all five architects received their participation fee. Once the winner was chosen, the five design proposals were publicized and the identity of the architects revealed. The winning architect was BIG, known worldwide for daring and unconventional architectural designs.

More than a loophole would be required, however, to achieve realization of this spectacular, but in terms of the competition, illegitimate architectural design for the new Maritime Museum.

Sidetracking legitimacy: insulating the instrumental concern from the legal one

The architect who refused the client's proposal convinced the Danish Association of Architectural Firms to take the client to court. The outcome of the court case was clear from the beginning. As even the client admitted, the competition was not conducted in accordance with the rules and regulations. The museum was fined DKK 150,000 or EUR 20,000), but the court did not rule that the competition was invalid for at least two reasons. One, the Association was not seeking a ruling of this nature. In the belief that quality of architecture was more salient than the economic interests of individual architects, strong voices in the field expressed the concern that the court would use the law to prevent a fine piece of architecture from being realized. To avoid a conflict, the Association only asked for a symbolic statement on the illegitimacy of the competition – which it got.

The second reason was more pragmatic. There was no longer a client that the court could hold publically accountable. In fact, the client and the sponsors had re-embedded the project institutionally. The foundations agreed to put their donations into a trust fund, the sole purpose of which was to act as client for building the museum. As a private business, it was not bound by the public sector's rules and regulations. As a result, a court could not order it to organize a new competition, allowing it to legitimately sign a contract with the architect behind the controversial design proposal. Furthermore, as a private business, the foundation could reap the available tax benefits to mitigate its investment in the new museum.

52 *Kristian Kreiner*

The museum leased the land from the local authorities for ninety-nine years, left the scene and allowed the trust fund to play the role of client. The museum reentered in 2013 as the happy recipient of the completed building that was offered as a gift by the trust fund.

In the process, the cost of the project increased by approximately 150 per cent (from DKK 130 m (EUR 17.5 m) to at least DKK 300 m (EUR 40 m) and the opening was delayed for several years. Nonetheless, nothing but praise and pride were voiced, this iconic museum almost instantly becoming a success, leaving only a few individuals wondering how to restore the legitimacy of public architectural competitions in the future.

Discussion

The case of the architectural competition for the new Maritime Museum of Denmark is unusual in at least two respects, one positive and one negative. Seldom do competitions realize their aim of ensuring high architectural quality to the extent that was the outcome of this competition. Competitions, however, seldom stir controversy to the point that clients are taken to court and penalized for not following the rules and regulations. The case description clearly shows that the client had little success in engineering a compromise, its choice plainly the prioritization of architectural creativity and quality. This resulted in the museum being indicted and penalized for being responsible for the illegitimate competition.

It could be argued that this merely illustrates the social acceptability of letting the end justify the means. However, the case unmistakably indicates the end need not be singular. Legitimacy, budgets, time and various other factors were apparently ignored. A peculiar aspect of this competition is the fact that the client was *able to make one goal* of such paramount importance that all other concerns were neglected.

What would have happened had the philanthropic foundations not so readily increased their financial commitment to the new museum? How could anyone have expected such a large budget overrun not to be held against the organizers and the client? Most often, clients protect their own legitimacy by insisting on receiving what they asked for in the planning phase and described in the competition brief. In this case, discovering what was actually possible to build caused the client and all those around them to change their minds. However, should they simply not have been more imaginative from the start and realized that the museum could be dug into the sidewalls of the dry dock, saving them from the subsequent embarrassment of having changed their minds? Such a solution was feasible after the competition, but not necessarily before the competition. The stipulated amount in the brief of DKK 130 m represented an estimate of what a new museum should legitimately cost, that is, it was an estimate of its worth. It seems unlikely that anybody could have convincingly argued why the new museum should cost DKK 300 m. Had they asked for such funding originally, the constraints set by the budget may have led them to be more realistic concerning their vision for the future museum. Thus, the restriction on the design task to fit the museum into the dry dock may reflect not only a limited imagination but also very real limits on the budget at the time of planning that would make digging the museum into the sidewalls of the dry dock impossible. The specific, spectacular design submitted by BIG forced everyone to see that spending additional money was worthwhile. In this case, it was the specific solution that drove the means, not the other way around.

Under ordinary conditions, the most likely trajectory would be that budget constraints would force the client to accept one of the more 'realistic' proposals. The fixed budget of DKK 130 m must have appeared adequate since it would have solved the housing needs of the Maritime Museum. Everybody would have been pleased and impressed with the foundations' generosity. By implication, DKK 300 m would have appeared excessive. An adequate budget would likely lead to an adequate museum, though not one that would land on the *New York Times*' coveted list of places to visit. The competition would have looked legitimate without sacrificing creativity and architectural quality as constraints spur creativity, the resulting level of quality relative to what is being asked.

A counterfactual history like this would most likely, by and large, be accepted, that is, everybody would deem the expectations to be realistic given general knowledge about budgets and institutional rules and values. However, alternative counterfactual histories are possible, even when deemed unrealistic. Institutional rules and values have no absolute predictive power because (a) occasionally they become neglected in the face of more important concerns and (b) because a completely integrated institutional field is an impossibility. The philanthropic foundations offered more money and changed the financial feasibility of the project. Important spokespersons for the architectural profession aired concerns that *architectural legitimacy* would be compromised if the illegitimately appointed winner was not accepted and awarded a contract. This perspective illustrates that the institutional field represents a complex and dynamic reality.

What made the institutional field change so radically as to make an illegitimate choice feasible and, in most other senses, legitimate? A good answer is probably the specific architectural design proposed by BIG, which clearly demonstrated what would be possible if the sidewalls of the dry dock could be built into. Everybody saw something more than a new Maritime Museum, although it was definitely also that. It was an architectural statement that apparently seduced everybody instantly. The jury was never in doubt, and the foundations were easy to convince that this design was worth giving extra funding. The press, the public, the profession, even the competitors, praised the architectural creativity and quality of the illegitimate design proposal. Its attractiveness was its incommensurability with the other design proposals. BIG's proposed design was in a different league, which, surprisingly, was not held against it. Most considered it *beyond* competition.

The specific design proposal motivated the foundations to exercise financial power and managerial shrewdness. Motivated to help realize BIG's spectacular design, they simply increased their donations and engineered the contracts to shield the project against any potential negative sanctions and prohibitions resulting from the court's ruling. In contrast to a previous case in which the legitimacy of a similar gift was challenged on the grounds of poor architectural quality, this case involved an outstanding piece of architecture that drove the flow of money, few expressing concerns about the legitimacy of the contractual retrofitting.

In conclusion, the fact that architectural creativity and quality prevailed must be ascribed to the financial muscle and managerial shrewdness of the foundations sponsoring the new museum. BIG's seductive design motivated them to extend their commitment, causing them not only to increase donations considerably but also to take responsibility for managing and realizing the project. In retrospect, the client might have wished to avoid the controversy by allowing a museum built into the sidewalls of the dry dock from the beginning. However, as indicated above, this would likely have been impossible due to the radically more complex engineering required and the concomitant budget demands that would have exceeded the notion of what a new museum should cost at the time. It was only when BIG's design proposal demonstrated what money could buy beyond the official target that the budget constraints were neutralized. Thus, it is hard to criticize the client and the organizers of the competition for lacking proper foresight and for defining the perimeter of the dry dock as the physical delimitation of the new museum. BIG's proposal forced the client and the architectural competition into a corner that did not exist a priori. For once, this corner proved the ideal place to be.

In a particular sense, the end justified the means in this case. Conventionally, the end is nothing but a promise of some future outcome and the choice of means perhaps unjustified if reality does not conform to the promise (Berger, 1974). The financial and professional power of the sponsors proved sufficient to ensure that the end materialized. Thus, the *happy* ending justified the means.

Managing dilemmas in ordinary competitions

The illegitimacy of the architectural competition examined here boils down to the simple fact that the rules of the game were settled after the game was over. This retrospective rule making enabled the jury to make a winner of the architecturally superior but legally unacceptable proposal. To retrofit the rules in order to favour a particular outcome is the epitome of unethical, unfair and illegitimate competitions. However, as we have argued elsewhere (Kreiner, 2012), such practices are not the exception to the rule but a necessary practice for accomplishing the task of the jury. Here, we will reiterate the argument as it applies to the competition currently under consideration.

The notion of a legitimate competition implies appointment of the winner without judgement on the part of the jury. However, the entries in an architectural competition are incommensurable and the jury's task is to make them commensurable. They do so by choosing a winning design and measuring the other designs against this winner. Such procedures are illegitimate in the sense that commensurability is constructed. Thus, an important and integral task of the jury is to make a practice like this that violates the notion of a fair procedure *appear* legitimate – a task which, in some ways, constrains which design to choose as the winner but does not accomplish selection of the winner alone.

As a result, the claim is that all competitions are illegitimate in the sense that the winner is selected by a different practice than the presumed one. In practice, and by necessity, the winner is selected before the criteria and rules are determined. The reason this procedure creates little controversy in most cases must be sought in the way competitions are managed. Most likely, competitions are managed to maintain the legitimacy of a procedure that cannot be legitimate. An image of legitimacy is fostered and protected by carefully drafted jury reports that evaluate each entry on its own merits, hiding the fact that the competition for primacy (March, 1999) necessarily implies comparative judgements. The winning translation of the future material building is chosen to represent the client's needs, thereby masquerading as the 'right' solution, which leaves plenty of room to criticize other translations for being partial, inaccurate or simply wrong concerning aspects the winning proposal differed on.

In the present case of the new Maritime Museum of Denmark, this image of legitimacy could not be maintained, not because it was illegitimate in any different sense than all other procedures for appointing a winner, but because its illegitimacy could not be managed legitimately. It is striking how little would have needed to be changed for this competition to have stirred no controversy at all. Had all the participating architects accepted the client's proposal, the competition would de facto have been cancelled, but the architects would have been compensated for their efforts as if the competition had been completed. However, this gesture on the part of the client may have signalled less than legitimate intentions. The client focused attention on the budget, claimed that all of the proposals violated the budget criteria, and then asked the architects to renounce all rights on the real issue, that is, the architectural solution space. While most of the architects resigned themselves to their fate and accepted the fee, one of them resurrected the competition, at least for a brief moment. Believing he had stayed within the budget, this architect undermined the client's attempt to end the competition legitimately. The architect's protest ignited the client's second contractual recovery attempt, which completely gave up pretending to be legitimate. The use of brute power was the fallback position that the client and the foundations chose to get their way, and legitimacy was the collateral damage of their strategy. They might have taken another position, namely to disqualify BIG and choose between the remaining designs. Apparently, the uniqueness of the BIG proposal made such a position highly unattractive and untenable.

In retrospect, however, the show of power seems to have been somewhat unnecessary. Had the foundations invented the strategy of recasting the role of the museum from client to beneficiary, a legitimate winner of the competition might have been appointed, while at the same time, the trust fund could legitimately have negotiated a contract with BIG.

The causes of the big controversy were very minor and had nothing to do with the competition brief, the composition of the jury, the quality of the design of the competition or the competence of the management. In the end, they boiled down to a question of *wanting* to maintain an image of legitimacy. Obviously, this competition might easily have ended as proof of the legitimacy of architectural competitions. It was only when legitimacy began to compete openly with the symbolism of power that legitimacy lost out. It would be tempting to use the museum that was made possible to justify this de-legitimization of the rules and procedures governing architectural competitions. However, as argued above, the competition might have found a legitimate winner without sacrificing the realization of the architecturally outstanding design. The fact that controversies concerning legitimacy manifested themselves in this case reflects an unusual lack of concern for maintaining trust in the architectural competition as an institution in society.

Conclusion

Replicating a discussion of substantive and procedural rationality in decision making (March, 1994), we may conclude that our case illustrates the potential conflict between substantive and procedural legitimacy. Substantive legitimacy (referred to above as architectural legitimacy) would suggest that the best design proposal should always win and be built. Procedural legitimacy would suggest that the best design proposal is identified by means of a preannounced procedure, including a set of evaluation criteria, ensuring, among other things, compliance with the competition brief. This is a legal and bureaucratic form of legitimacy. The ways in which architectural competitions are organized, staged, framed and managed are meant to ensure a link between substantive and procedural legitimacy. Most of the time, there are good reasons for arguing that a procedurally legitimate competition will lead to a substantively legitimate outcome. This, however, may build on rationalization, not calculation, since, for example, criteria like 'The overall architectural, aesthetic, functional and technical quality of the design' are inherently ambiguous but acquire meaning and clarity in the jury's process of selecting the winner, where the winning design proposal is attributed the requested qualities. Such attributions explain the outcome but do not cause it (Kreiner, 2012; Rosenzweig, 2014). Because of such rationalization practices, the two types of legitimacy seldom appear to be in conflict, making legitimacy a non-issue in spite of the fact that competitions are routinely concluded in a different manner than the presumed and procedurally legitimate one.

When, as in the present case, legitimacy is an issue, we suggest it is because the winning design proposal defied rationalization. There was no way of arguing that the winning design complied with the brief as it undeniably encroached upon both physical and budgetary delimitations. Therefore, the prescribed procedure had to be openly violated. Had the proposal not been so compelling, also in the eyes of the foundations, it would have been disqualified as a matter of routine. But, according to the jury and its sponsors, the proposal was too good to be disqualified, creating an undeniable rift between the substantively and procedurally legitimate winners. There may have been many reasons behind the blunt show of power, but, in the current context, it translated into a concern for *substantive* legitimacy. The funders could afford to pay for the best design proposal and were not willing to compromise merely because it violated the competition brief. Few individuals or institutions are able to parade such power, which is why most clients try to find a shrewder way of catering to substantive *and* procedural legitimacy. Few possess the financial muscle to raise hundreds of millions of Danish kroner to build an illegitimate entry in an architectural competition. Few juries face a situation in which the competition successfully produces an outlier in terms of architectural quality that defies ordinary rationalization. The combination of conditions for this particular competition was highly unusual, but when mixed with the ability and willingness to demonstrate power, the competition between creativity and legitimacy was made manifest and visible to all.

Architectural competitions are organized and managed to help the participating architects produce a creative design proposal and to ensure that the winning design proposal is in compliance with the rules of the competition, as stipulated in the competition brief. We argued that competition between these two concerns, that is, creativity and legitimacy, is an inherent part of architectural competitions. In order to explaining why, in ordinary cases, these concerns seem to coexist peacefully, we suggested that rationalization plays an important role. The proposal that the jury – for whatever reasons – finds the best entry is rationalized to be the legitimate interpretation of the competition brief. Other design proposals building on alternative interpretations are devalued or even disqualified. The *substantively legitimate* winner in the eyes of the jury also transpires as the *procedurally legitimate* winner. In extreme cases – like the competition studied here – the winning design proposal is simply too much of an outlier to be rationalized as a legitimate interpretation of the competition brief. But, as elaborated upon above, it takes unusual conditions for such an outlier to reveal the inherent competition between creativity and legitimacy. Had the foundations been unwilling to increase their donations significantly in the interest of getting the outlier design built, it would have been disqualified and one of the other proposals rationalized as the most creative interpretation of the competition brief. Nobody would have questioned the legitimacy of disqualifying such a deviant proposal, claiming that the architect's creativity was wasted on solving a different task than the one defined. The reality, however, was that the foundations increased their funding and the outlier was built in the end. Therefore, the competition between creativity and legitimacy became manifest.

Finally, does the competition studied suggest that architectural competitions generally deliver the intended effects that constitute them as a social mechanism? In spite of all the controversy, the answer has to be affirmative. We saw how the competition enabled creativity in the design of the new museum in a way that alternative procurement forms could not. A competition allows a number of isolated teams to work in parallel on their own set of ideas and interpretation of the brief. It produces a situation in which each team can go off on a tangent in search of an original but consistent design proposal. The outcome is not only ideas but translations of such ideas into concrete designs, allowing the jury to evaluate not the ideas but the proposals. Imagine if the client had been able to pick one architect to work with from the beginning. With five contestants, the chance of picking the one with, in retrospect, the best ideas, would appear to be one in five. If only one architect had been picked, the design development process would likely appear to be stuck on a suboptimal trajectory from the start. Perhaps even more important, the interaction between client and architect would likely be continuous, allowing the client to give feedback on ideas and the interpretation before they had the space to develop into design proposals. The client's clarification of the architects' questions early on does, in fact represent feedback on the idea of building the museum into the walls of the dry dock, reiterating the importance of respecting the perimeter of the dock. Had the client had the opportunity to follow the work of the individual architectural team, they would most likely have been directed back to what, at the time, appeared to be a proper and realistic design path. The opportunity for being positively taken by surprise in the end would have been forsaken. Thus, the creative effects of the competition were facilitated by the isolation that the teams enjoyed in preparing their proposal, that is, isolation from each other but, most importantly, also from the client.

Indirectly, we also saw how the competition fostered legitimacy. Again, isolation plays a role, but this time it is the isolation of the jury that proves functional, allowing it to rationalize that the best proposal was also the correct interpretation of the brief and, consequently, the legitimate winner. Such a practice does not reflect foul play or bad intentions on the part of the jury but occurs out of necessity, the competing proposals being fundamentally incommensurable. Making them commensurable requires new rules of the game, not least in terms of reinterpretation and prioritization of the brief's requirements and the client's preferences. When legitimacy becomes an issue, it is likely the result of this isolation being broken. In the present case, the jury had to involve the participating architects because the winning proposal was an outlier impossible to rationalize as a proper interpretation of the brief. While the jury almost succeeded in maintaining legitimacy, it was lost when one architect refused to accept the deal offered. Still, the case illustrates how unusual conditions must prevail before juries lose the ability to claim that the best proposal is also the procedurally legitimate one. The conventional organization of architectural competitions successfully hides the fact that the contestants do not merely compete to produce the best solution to a given task but also compete for the most compelling interpretation of the task. The outcome of such competitions cannot be decided without some form of value judgement on the part of the jury as to which proposal is the most compelling. The winning proposal can never be found simply by following a formula if creativity plays an important role. The practical organization and management of architectural competitions enables juries to maintain procedural legitimacy, even when facilitating and rewarding creativity that inevitably seems to undermine it.

We would like to conclude with a word of caution. It is worthwhile to celebrate the architectural competition for its capacity to make a winner of the best architectural solution and for producing legitimacy by rationalizing the solution as the correct translation of the competition brief. We saw that there are limits to the ability of juries to rationalize the outcome as legitimate but that such limits are quite broad. However, the power to rationalize the best architectural solution can be exercised equally well for less noble purposes. In fact, questioning the legitimacy of most winning designs, excellent or mediocre, is difficult. When, in ordinary cases, the competition produces alternatives that do not differ radically, the outcome becomes unpredictable but also less consequential. The freedom to rationalize any outcome chosen nevertheless becomes instrumentally important in bringing the competition to a peaceful conclusion.

Note

1 The distribution of roles is unclear. The jury and the philanthropic foundations played an active role but often on behalf of – and in the name of – the client. When we ascribe agency to the client, this wider coalition of actors is implied.

References

Berger, P. L. (1974). *Pyramids of Sacrifice: Political Ethics and Social Change*. New York: Basic Books.
Hedström, P., and Swedberg, R. (1998). Social Mechanisms: An Introductory Essay. In P. Hedström and R. Swedberg (Eds), *Social Mechanisms. An Analytical Approach to Social Theory* (pp. 1–31). Cambridge: Cambridge University Press.
Kahneman, D. (2011). *Thinking, Fast and Slow*. London: Allen Lane.
Kreiner, K. (2012). Organizational Decision Mechanisms in an Architectural Competition. In A. Lomi and J. R. Harrison (Eds), *The Garbage Can Model of Organizational Choice: Looking Forward at Forty (Research in the Sociology of Organizations, Volume 36)* (pp. 399–429). Bingley, UK: Emerald Group Publishing.
Kreiner, K. (2013). Constructing the Client in Architectural Competitions: An Ethnographic Study of Architects' Practices and the Strategies They Reveal. In M. Rönn and G. Bloxham Zettersten (Eds), *Architectural Competitions: Histories and Practice* (pp. 217–245). Fjällbacka, Sweden: Rio Kulturkooperativ.
March, J. G. (1994). *A Primer on Decision Making. How Decisions Happen*. New York: Free Press.
March, J. G. (1999). Exploration and Exploitation in Organizational Learning. In J. G. March (Ed.), *The Pursuit of Organizational Intelligence* (pp. 100–113). Oxford: Blackwell.
Rosenzweig, P. (2014). *The Halo Effect … and the Eight Other Business Delusions That Deceive Managers*. London: Simon & Schuster.

3 The best of both worlds?
Client decision making in architect selection processes

Leentje Volker

Introduction

Architect selection processes are interesting phenomena. As pointed out by Strong (1996) the potentially conflicting issues in design competitions originate from the diverse roots of the phenomenon: the design competition, the tendering for works and services, and the search for a design partner. A design competition is organized at a very early stage of a construction project as a first connection between acquiring suitable accommodation and hiring a designer to create a representation of the building. Although the service delivered by the architect cannot be directly related to the actual product delivery and use of the building, they are connected in the minds of the decision makers. Among these deliberations decision makers also have to comply with (inter)national rules and regulations, such a European procurement law, government policy and sustainability issues. Hence, selecting the right architect requires different kinds of decision strategies and domain specific skills of decision makers among commissioning bodies (Volker, 2010). In this chapter design competitions are considered selection processes of clients for an architect who offers design solutions for their accommodation problem. Hence, since the study focuses on European public clients in particular, EU procurement law applies to these selection processes. This leads to two worlds of justification: the professional world and the civic world. In this chapter I will identify different decision strategies on dealing with these potentially conflicting rationales and propose elements for the design of a tender process which takes these processes into account.

The underlying logic of EU procurement law is that an open procurement market and free movement of goods and services in the European Union would ultimately benefit all citizens. Public procurement regulations aim at safeguarding business connections between government and market parties to create 'best value for taxpayers' money'. The sums of money involved in procurement are substantial: in an evaluation the EU reports over 150,000 invitations to tender in 2009 in conformity with EU Directives by 35,000 authorities with a total value of €420 billion per year, driving down costs by around 4–5 per cent and generating savings of approximately EUR 20 billion (Internal Market and Services Directorate General of the European Commission, 2011). In the case of design competitions under EU procurement law, the client organization is considered a contracting authority, announcing the award of a contract to an architectural firm as a provider of design services. It is represented by employees or other actors that actually make the decisions and judge the quality of the design proposals during the tendering process.

The client organization plays a crucial part in preparing the legal connections between an architect and a contracting authority by designing the selection processes and implementing the relational and contractual agreements that follow (Hartmann, Davies and Frederiksen, 2010). Decision makers involved in procuring architectural services face the need to justify what they do towards public opinion as well as the professional expectations in the architectural community that either joined the competition or debate about the outcomes. This exemplifies a clash between the 'world of fame', the 'inspired world', and the 'civic world' (Boltanski and Thévenot, 2006). In the context of this research we consider the professional world as a combination of the world of fame and the inspired world. The world of fame refers to peer discussions and professional reputation among architects. The inspired world is considered as visionary, inspirational and creative, delivering design solutions to realize dreams and ambitions of client organizations. For architects, being published in magazines and winning design prizes is essential to being acknowledged as a professional. Traditionally design competitions open up opportunities to expand their portfolio and run successful businesses (Manzoni, 2014). Since 'public' implies democracy and open debate on preferences, this often compels client organizations to actively involve their stakeholders and architect to convince their future users of the potential value of their propositions.

Next to the professional world stand the laws and regulations of procurement safeguarding the free movement of goods and services, in other words, the procedural justice of the civic world serving the general interests. These general interests concern the quality of the built environment as defined by the users and other stakeholders. Since we all live and work in the built environment, decision makers often compromise between private and professional interests: for example, a mayor is responsible for making a decision for a new city hall that fits the needs of all citizens in a municipality but will also be a future user of the building himself. Public procurement regulations consider tenders as managerial processes in which a rational choice between alternatives needs to be made in a process that seems free of political context and interrelated stakeholder interests (Harrison, 1999). Prior announcement of decision criteria and decision methods would help to instantiate the EU principles of equal treatment, transparency, objectivity and proportionality, and inform participating companies what to expect. Hence, client organizations have to announce their ambitions at the beginning of the selection process, accompanied by a description of the exact procedures that enable decision making on the best designer.

Procurement decisions for construction projects are, however, often taken from a multiple stakeholder perspective in a political environment. These decisions are further complicated by the fact that they are based on sketches, models and drawings of the proposed building – visual representations of the future product rather than the product itself. This entails that decision and justification processes are often incremental in order to create decision support (Hodgkinson and Starbuck, 2008). In many cases, politicians, board members, citizens, government employees, professionals and other stakeholders play a role as decision maker in a complex political arena aiming at making the best decision for the project. Due to the complexity and (unforeseen) dynamics, sensemaking is essential to give meaning to the complicated decision task which entails much missing information on the actual impact of the decision (Maitlis, 2005).

The process of sensemaking involves the retrospective development of plausible meaning that rationalizes what people are doing (Maitlis and Sonenshein, 2010;

Weick, 1995). This work connects sensemaking to the process by which decision makers negotiate the different economies of worth (Boltanski and Thévenot, 2006) and tries to conform to the expectations of a rational decision making process, which has recently gained interest (March, 2006). The majority of psychological studies have questioned the capability of humans to make rational decisions and have instead emphasized the incremental and political character of decision making, and the importance of sensemaking, expertise and intuition (Beach and Connolly, 2005; Hodgkinson and Starbuck, 2008). Yet, rationality is still regarded as the 'almost universal format for justification and interpretation of action and for the development of a set of procedures that are accepted as appropriate for organisations pursuing intelligence' (March, 2006, p. 202).

The aim of this chapter is to address the decision making process during the awarding decision for an architect in the context of EU procurement regulations. This decision announces the start of a collaborative process between a governmental client organization and an architect. Especially during the procurement of architectural services for a public building the impact of these decisions is substantial and the clash of different forms of justification is inevitable. My research tries to find out where the current conflicts arise and how actors deal with this. The study has a similar ethnographic approach to the work of Gkeredakis, Swan, Nicolini and Scarbrough (2011), Langfeldt (2001), and Kazemian and Rönn (2009) who studied funding decisions of expert panels in different fields. Yet, these studies did not include the rationale of the legal procurement framework and the pressure of public justification to a diversity of public stakeholders.

In this chapter, first an explanation is provided on the character of making decisions about selecting architects and the literature on sensemaking. Then the research approach and set-up of a comparative study of four tender cases is described. In the results the importance of sensemaking in expert based decisions in procurement situations is highlighted. Finally, the implications for future design competitions are discussed. The chapter is therefore of interest for scholars in organizational decision making, project management and public procurement, but also for administrators, consultants and managers in the public sector.

Making decisions on architectural design

The decision to select an architect for a public building is characterized by surprise, dialogue and ambiguity (Kazemian and Rönn, 2009; Kreiner, Jacobsen and Jensen, 2011), the allocation of large public funds (Strong, 1996), and the involvement of numerous stakeholders (Volker et al., 2008; White, 2014). Architectural design is a professional skill based on education and experience gained through practice (Mieg, 2008). Traditionally, architectural design services have been procured by design competitions in which an independent jury panel, sometimes including one or two representative of the client body, evaluates the proposals that have been submitted (Strong, 1996). Research on designers and comparable professionals has shown that experienced practitioners interpret and manage complex and demanding situations faster and more accurately by using tacit memory schemes (Rosen et al., 2008). Their domain relevant experience enables them to make intuitive decisions based on their tacit knowledge and unconscious memory. Members of the same profession share this code, and will accept peer review from within their discipline (Mieg, 2008).

Based on publications that describe the jury processes (e.g. Kazemian and Rönn, 2009; Kreiner, 2007) (see also Kreiner and Silberberger in this volume), it can be concluded that pattern recognition (Tversky and Kahneman, 1981) takes place. In architectural competitions judgement tasks are complex and involve moral, ethical and aesthetic dimensions. Only limited time and information is available and there is social pressure. Therefore decision makers have to respect the different views of the panel members and be aware of the setting in which they take decisions (Lamont, Mallard, and Guetzkow, 2006). Gkeredakis et al. (2011) found that deliberation practices in making healthcare funding decisions constitute an assemblage of three interrelated activities: performing procedural requirements, making sense of decision cases and deliberating the merits of cases. Ranking methods have a substantial impact on the outcome of the review (Langfeldt, 2001).

In all previous research settings the actors are domain specific experts (peers and professionals) making decisions as part of routine procedures and daily work to judge proposals and distribute (public) money. However, many building projects such as the realization of a municipal office, a new or adjustment to theatre, museum or library are one-size, unique objects realized in a temporary interdisciplinary project context. Only a few of the awarding authorities can be considered professional clients managing a real estate portfolio in a certain sector, and thus be expected to be experienced in architect selection processes. This implies that the governmental organizations that deal with procurement situations for new buildings often do not have a routine and lack the required domain-specific expertise. Yet, they still face similar justification issues (Boltanski and Thévenot, 2006).

Sensemaking in organizations

Decision making and organizational sensemaking are closely related because 'decision making stimulates the surprises and confusion that create occasions for sensemaking' (Maitlis, 2005, p. 21). The process of sensemaking involves the retrospective development of plausible meaning that rationalizes what people are doing (Maitlis and Sonenshein, 2010; Weick, 1995). As Weick and others such as Maitlis (2005) have pointed out, sensemaking is particularly critical in dynamic and turbulent situations where the need to create and maintain coherent and common understanding is important. People then produce or reactivate accounts to deal with uncertainty and ambiguity and include these in their mental models in order to make decisions. According to my research perspective, these situations can be compared to architect selection processes organized by incidental client organizations – either due to the lack of frequency or by complexity of the construction project.

Balogun, Pye and Hodgkinson (2008) conceptualize sensemaking as a social process of constructing and reconstructing meaning, which enables individuals to collectively create, maintain and interpret the world through interacting with others. The intertwined concepts of 'framing' (shaping the meaning of a subject and sharing it with others), 'sensegiving' (attempts to influence sensemaking and construction of meaning toward a preferred redefinition of social reality), 'sensereading' (perception of circumstances and aligning of interpretations), and 'sensewrighting' (inheriting, shaping and reflecting the understanding of the world) are all related to the resource, process and meaning of power effects in organizational decision making (Balogun et al., 2008). Gioia and Chittipeddi (1991) found that change processes displayed a sequential and reciprocal cycle of sensemaking and sensegiving to expanding audiences, such as the CEO, top managers, organizational membership groups and other stakeholders.

According to Maitlis (2005) social processes of sensemaking among large groups of diverse organizational stakeholders are relatively under-researched compared to the cognitive aspects of sensemaking or social processes in crisis situations and extreme conditions. Hence I take up the call of Balogun, Pye and Hodgkinson (2008) for more research from the perspective of naturalistic decision making, which focuses on making sense of deciding and is shown in situations in which 'the whole spectrum of calculability situations located between a "purely objective" calculability and a "purely subjective" judgement' (Cabantous, Gond and Johnson-Cramer, 2010, p. 1556) is applied. This also makes it possible to contribute to the theoretical embedding of the naturalistic decision making tradition in existing cognitive, social and organizational theory, as advocated by Beach and Connolly (2005), and to complement existing views on how rationality unfolds in organizations.

In this context I am especially interested in the different kinds of sensemaking processes in high-stake situations. Architect selections can be considered as high-stake environments because of their political sensitivity, the large sums of public money spent and the high impact of the built environment on citizens' wellbeing. Building projects have a high political impact outside the usual domain of politicians and civil servants. They are also rare events for most of the decision makers – situations in which sensemaking could lead to organizational learning (Christianson et al., 2009). The research illustrates the complexity of multi-way sensemaking processes between employees and other lower stakeholder groups, middle managers, top managers and politicians.

Research design and methods

To study how processes unfold over time, longitudinal data collection from multiple data sources is desirable (Langley et al., 2013). Four rich and exemplary case studies of public authorities selecting architects for their future housing solutions are used. In each case, project teams were appointed by the client organization to organize the tender competitions, consisting of employees of both the purchasing department and the real estate department, consultants, and other key stakeholders such as board members or other administrators. I followed each project team for five to twelve months collecting observations, interviews and archival data over a period of three years. This cross-case replication allows testing and deepening of theoretical ideas in different settings, leading towards a theoretical model on different navigating strategies (Flyvbjerg, 2004; Langley, 1999). In the analysis the organizational world is seen as socially constructed and people as knowledgeable agents (see also Gioia, Corley and Hamilton, 2013). By using the perspectives of official commissioners, project team members, architects and other stakeholders in different episodes and by combining in vivo observations, memories, interpretations and artefacts, the micro-processes of interaction and interpretation from different perspectives are constructed.

Although freedom of information and transparency of public governance would imply otherwise, gaining access to tender situations proved to be very difficult. Tender situations can be sensitive and delicate, and scouting tenders before their official

64 *Leentje Volker*

announcement to investigate the processes leading up to a tender was virtually impossible. Because this research aimed at developing theory instead of testing it, theoretical sampling is appropriate. In this situation cases were selected because of opportunities for unusual research access and revelatory situations.

The first three instrumental cases to be investigated conformed to the rules of a restricted tendering procedure: a School, a City Hall and a Provincial Government Office. Additionally one case about an ideas competition for a new Faculty Building of a university was investigated. In almost all cases international architects were involved as participants, despite the fact that all clients were Dutch. Table 3.1 shows the representative of the commissioning client body and the other actors involved in the decision process. In all cases architects were somehow involved in the selection process; sometimes as part of the jury (City Hall, Faculty Building), in other cases as consultant (School, Provincial Office). The cases show the process of architect selection from a psychological perspective in their full complexity including the interrelations of all phases, actors and characteristics but differed in the scope of the brief, the type of tender, the actors that were involved, and the characteristics of the selection process (see Table 3.1). We characterized the three tender cases as pragmatic, democratic and political in terms of their potential fit of the aims with the participation strategy and design of the tender procedure. The case of the Faculty Building is characterized as competition (see Table 3.2).

All tender cases included the restricted procedure with economically most advantageous tender criteria. This is the second most popular EU procedure after the open procure, and costs much more in time and expenditure than other kinds of procedure due to the high value and complexity of the contracts. A restricted tender procedure is similar to a two-stage design competition and consists of three phases: 1) a selection phase in which the potential candidates are screened and selected based on a first announcement of the problem; 2) a tender phase in which the architects interpret the problem and prepare proposals to show their vision of the contract task; and 3) an award phase in which the most suitable candidate for the job is selected after a process of organizational decision making.

In the period of 2006–2008 a variety of different forms of data for each case were collected for triangulation between self-report, observed behaviour and official justifications (see Table 3.1). The observations were based on field notes taken during the decision process throughout the selection process, so before, during and after the final decision making of the jury. In the pragmatic, political and competition cases I got involved before the announcement of the competition, which allowed me to observe the decisions made to prepare the documents as well as the implementation of them, both at the level of the project team and the steering board. In the democratic case I became involved just after the first round of selection, enabling observations after the selection phase only. In all cases I attended most of the official meetings in which the jury panel discussed the alternatives. In the political and competition cases I was also involved as a participant in the preparations of the final tender process. Sometimes these meetings took place in a regular meeting room, sometimes they were in the location of the client (e.g. a class room, theatre or city hall), always enabling me to note down everything said by the decision makers from a close distance. In order to gain an overall overview of the case and collect different perspective on the processes, retrospective semi-structured interviews were conducted with key decision makers and members of the project team after the tender decision had been officially published

Table 3.1 Overview of the case characteristics.

Case	Commissioning client body	Actors involved in decision making	Artefacts for selection	Case data
Pragmatic: School with Sports Facility	School board and municipal department of sports, representing municipal department of education	Members of school board, employees of the municipal dept. of sports and dept. of education, representatives of the local community	Portfolio, reference projects, CV leading architect, presentation of design vision	14 hrs non-participatory observation; 8 semi-structured interviews; 8 documents
Democratic: City Hall with Library	Board of mayor and alderman with library representatives, representatives of city council	Representatives of all political parties, advised by citizens, experts and user groups	Reference projects, profile of leading architect, sketch design, scale model, presentations	7 hrs non-participatory observation; 9 semi-structured interviews; 6 documents
Political: Provincial Government Office	Provincial executives and queen's commissioner, representatives of provincial council	Members of the executive board, head of the departments of provincial organization	Concept design and presentation	53 hrs non-participatory observation; informal conversations; 15 documents
Competition: Faculty Building	Dean of the faculty of architecture, representatives of university board	Chief government architect, two internal architecture professors, two international architecture professors, MSc student, director of the architecture museum	Design sketches on one or two A1 format posters	32 weeks of participatory observation including jury meeting; 6 semi-structured interviews; 5 official documents & numerous informal documents

(see Table 3.1). Together with the official and preparatory documents as gathered during the selection process, this improved the understanding of multiple perspectives on the processes.

The data were initially analysed as separate case identities, and then systematically compared on appearing constructs collaboratively as a research team in Atlas. ti, a software package to support qualitative coding. Throughout data analysis and reporting we (my research team members and I) frequently went back and forth between the interpretation, original data and theoretical insights. In all phases of the decision process we realized that making a decision on the right architect requires a continuous consideration of demand (the programme of requirements for the project) and supply (the offered design solutions). We also saw that an initial decision needs be checked against the expectations among the members of the jury committee in order to be officially announced and supported by the community and the other participants in the competition. We therefore distinguished between making sense of the decision task, the phases of a decision making process and justifying the decision. These processes are iterative and thus continue throughout the case. At the same time, decision makers have to acknowledge that it is hard to draw up a decision process without knowing the decision task, and a justification is more complicated – or impossible – without a clear decision task and decision process. The differences between logics of justification between the professional and the civic world were visible in conflicts in the system rationale (perspective on the tender), process approach (the actual managerial decisions) and decision outcome (the selected architect). During further analysis the way in which the actors dealt with these conflicting logics as sensemaking processes was theorized. As a means of confirmative validation, a workshop was conducted with eight domain experts who had experience as clients, jury members and submitting architects as well as expertise in procurement law (see also Volker, 2010). The findings with respect to the different worlds of justification are integrated in the description of how actors dealt with the frictions in adopting a sensemaking approach. First an overview of the processes is given, then the specific sensemaking processes are expounded.

Results on client sensemaking processes

From the perspective of the client, the tender processes were high stake situations of organizational decision making for those involved. They also experienced the clash of different economies of worth: the functional requirements for the building from the inspirational world met in Boltanski and Thévenot's (2006) terms questions of the right procedure from the civic world and matters of reputation from the professional world. Although the tendering law was intended to promote market principles in the realm of public spending, it manifested itself mainly as procedures; the market perspective of value maximization had been transformed into a set of rules for decision makers acting on behalf of the public instead of individual self-interest.

Despite the clashes of rationalities, decision makers apparently found a way to deal with this interplay of logics. This is theorized as sensemaking and distinguish several forms of sensemaking based on a combination of the concepts by Balogun et al. (2008), the elements of effective investment decisions of Butler et al. (1993), the matching concept of heuristics and image theory, and the different decision rationalities as described by Simon (1997), Boltanski and Thévenot (2006), and Miller and Wilson (2006). These sensemaking activities are:

1. *Reading the decision task* – What are the aims and opportunities of the selection process?

    The sensemaking process of reading the decision task is based on the concepts of sensereading and framing as described by Balogun et al. (2008) and Beach and Connolly (2005). It deals with the translation of the aims of the client into a tender procedure for the selection of an architect (Jones and Livne-Tarandach, 2008), which requires colloquial and informal communication (Gkeredakis et al., 2011). The development of a tender brief (e.g. we want an iconic building that integrates with the surrounding urban structure) and the analysis of the project environment (e.g. who will be the main users?) are important parts of this sensemaking process.
2. *Writing the decision process* – How is the decision process 'designed' in order to meet the demands of the procedural and the professional world?

    This process is based on the concepts of sensewriting, sensegiving and framing as described by Balogun et al. (2008), Gioia and Chittipeddi (1991), and Beach and Connolly (2005). This process entails the writing (shaping) of the selection process of an architect by the client during a project. Upfront, decisions need to be made about when and how decisions are taken and which kind of information is required to do so in a transparent and objective manner. The level of expertise of the decision makers appeared to be an important factor of influence in the decision process. An example of an element in the writing process is the composition of the jury.
3. *Justifying a decision against different economies of worth* – How is the decision justified against different economies during jury board meetings and after the decision had been taken?

    This third sensemaking process was needed to feed the rhetoric of justification (Cabantous and Gond, 2011). It deals with the explicit justification of the decision at the end of the process against the different economies of worth (Boltanski and Thévenot, 2006) that are present during the selection process for an architect. The representative of a client has to justify their final decision to their own organization, to the public, to society, and to the architects that joined the tender. These multiple responsibilities are described by Thompson (1980) as the many hands that make it difficult to identify one single person responsible for a tender decision.

These three sensemaking processes are strongly interrelated but also follow a certain sequential order. Table 3.2 shows the most essential findings of the phenomena as found in the cases. In the next section the concepts are explained from the literature and the empirical results of the four case studies.

Reading the decision task

In all cases the tender enabled clients to reach multiple aims that were often more strategic in nature than awarding a contract would imply. This multitude of aims was usually reflected in the arguments that were used to justify the decision for a

Table 3.2 Overview of findings per case in relation to sensemaking processes.

Case characteristics/ Sensemaking process	Pragmatic	Democratic	Political	Competition
Reading the decision task	– Searching for a project partner with relevant experience – Click with architect is most important – Award decision based on presentation of architect, no plan required	– Searching for a plan of a secured partner – Click with design proposal is most important – Domain specific experts involved in advisory committee	– Searching for budget secure alternative – Political support more important than building itself – Awarding decision based on building specifications and scale model	– Searching for inspiration – Design quality can be recognized by experts based on a visual representation – Anonymous designs
Writing the decision process	– Jury panel in selection and awarding with similar parties – Decision authorities differed per phase – Domain specific experts in jury panel	– Low profile selection decision, award decision fully democratic – Stakeholder groups involved in advisory role – Scale model required in award phase – Extended interaction with designer	– Project team, steering committee and jury panel strongly interwoven – Stakeholders not actively involved – Expert only involved as 'educating' advisor →Cancellation of decision process after selection phase*	– Consultation with support of expert panel – Authority of jury panel of domain specific experts – Participation options limited to participation in competition, not in assessment
Justifying the decision against different economies of worth	– Motivation of decision in matrix scheme with criteria – Open to questions – Legal non-compliance term	– Press release with jury statement – Official letter with short motivation – Open to questions – Legal non-compliance term	– Legal minimum left after exclusion grounds so not application of selection criteria – Open to questions – Motivation in matrix scheme with criteria and short motivation	– Jury report for architectural community – Active PR policy for 'outside' – Not open to questions, jury members restricted to jury report

* Because the case was cancelled after the selection phase, final decision making based on design quality did not take place.

winning architect but not always made explicit in the tender design. At the same time it was found that the intertwining traditions make the need for reading the decision task more urgent. It was found that the most important dilemma which clients faced in the reading process was a distinction between the search for 'a click with the design', as suggested by the tradition of design competitions and 'a click with the architect' as a partner delivering services, as suggested by the tender principles. In the democratic case and the competition case, decisions were framed as a search for a design. The design clearly acted as a boundary object in decision making (Bresnen, 2010), which was not the case in the pragmatic case. In this case the most dominant frame was *'the search for the right partner'* capable of designing the future building and *'realizing their dream'*. This was also reflected in the criteria that this client applied during the decision process. The perception of circumstances and development of the decision frame can therefore be considered as essential parts of the sensereading process.

For the client, the selection of an architect constituted an interactive search for a person or group who could visualize and implement their needs and ambitions best. The decision making process starts when representatives of the organization articulate a housing problem or the political wish for a representational building. The information on which the project definitions are usually based often become obsolete by the time a judgement has to be made. This makes the identification of decision criteria and allocation of weights to the criteria more complex than is presumed in the logic of procurement law. John, one of the architects in the democratic case, characterized this process in a retrospective interview as follows:

> *Well, the city council needed a new City Hall. This means the council needs to develop a feel for what they want. And eventually with that feeling they fell for the winning party. Subsequently they have to substantiate that. [...] So in the end you'll see that the winning submission only has positive features. We have been involved in so many tender processes; we know both sides of the story.*

The logic of procurement entails that the brief describes the aims of the project in the official announcement. If the brief is to support the decision process, its level of detail should be aligned with the aim of the tender and the proposed procedure. Since a sketch design is made at a later stage in a building project, it requires a different kind of tender brief with more detailed functional, financial and contextual information than a tender brief for a visionary proposal for a potential design project, which should first meet the political and social aims in order to enable a decision about the design strategy.

However, the problem with announcing requirements in the brief is that over time a client's requirements may change. This happened in the competition case. The competition brief was set up several months after a fire had destroyed the previous faculty building. During the duration of the competition project the faculty had found shelter in an old office building. By the time the jury judged the anonymous entries to decide about the winner – the best solution to their housing problem – this temporary accommodation had received a warm welcome from the employees, the executive board members and the architectural community. The jury decided to award two of the six prizes to entries that proposed transforming the temporary housing into a permanent one. The change in context also altered the scope and the brief of the project. Reading the decision task can therefore be considered as an ongoing activity that requires attention throughout the tender process. However, the logic and routines of tender procedures do not allow for such adjustments during the process.

Uncertainties developed that were related to having no influence on or control over the decision alternatives as they were being developed by submitting architects. According to the law, the only opportunities a client has for controlling the quality of the service lie before announcing the tender and during the evaluation of the alternatives. In the cases studies it was found that the selection process was a result of the decision makers' interaction with the alternatives once they were confronted with them and began to make sense of the proposed designs. This interaction between the decision makers and the design alternatives has a firm place in the competition tradition (Kazemian and Rönn, 2009). Procurement law, however, assumes that the procedure and criteria can and must be designed up front, which in our sample clients found it almost impossible to do. Instead, the process showed similarities with the unfolding iterative cycle of Gkeredakis et al. (2011): performing procedural requirements (mainly before and after jury deliberation), making sense of cases (in different rounds and by different stakeholder groups), and deliberating funding merits (re-interpreting an offer, articulating, sharing and debating arguments in order to formulate a rational decision). This implies that tender processes should provide room for such iterative processes.

Writing the decision process

Regarding the writing of the decision process, it was found that on a general level the findings of the execution phase resembled the six stages of the selection process described by Kazemian and Rönn (2009): submission check, determination of order of work, choice and preliminary judgements, presentation of interesting contributions, ranking, and decision making with architectural criticism. Being involved as participatory observer in the competition case, it was possible to influence the structure of this jury meeting and validate this structure by explicitly designing the jury evaluation process. Based on these experiences and the observations in the pragmatic and political case, it became clear that in every phase of the decision process, a group of decision makers went through their own writing process of sensemaking. During the different phases of the decision process the jury members were affected by external influences, such as the opinions of other members of a group, changes in the context such as time pressure, or personal factors, such as moods and emotions. Consequently, decision making was a dynamic, incremental and cyclic process based on several kinds of value judgements.

It was usually during the judgement phase that decision makers started the development of a frame of reference based on the aim of the assessment and the documents to be assessed. This implies that preparations for an assessment frame can be made, but the definite assessment frame cannot be developed in the absence of the actual submissions. In this sense, assessment could be compared to the process of qualitative data analysis: the structure of analysis arises from the data. This is contrary to most quantitative research that typically consists of testing assumptions that have already guided the process of collecting data. Procurement law assumes a quantitative process, while our findings suggest a qualitative process. In all cases the decision criteria were somehow used to build a frame of reference between the stakeholders.

In all cases members of a jury panel reached a consensus by several rounds of ranking, discussion and/or voting, similar to what Langfeldt (2001) describes in her work on grants assessments. The consensus proved not to be the same as an average of opinion but rather the result of a negotiation process, which is in line with the results of Kazemian and Rönn (2009) on jury processes in design competitions. Decision makers needed time to interpret the criteria, the assignment and the brief that was mostly built by others not belonging to the jury panel. The observations confirm that experts were better at seeing the significance of information, identifying important cues for risks, estimating consequences and judging autonomously (Hutton and Klein, 1999). Experts also felt the need to discuss and harmonize their preferences with other members of the group, which contributes to legitimization of the decision to the participants and society. Further analysis of the democratic case showed that the experts addressed more aspects of design quality than the user groups and citizens (Volker et al., 2008).

The more expertise is available, the more strategic a decision process can be. In the pragmatic case Ewan, an urban planner by background, used strategic voting to combat the logic of the tender in the first round of selections:

> *What I did then was, for the submissions that positively surprised me and which I thought should have an important role in this selection, I strongly favoured them by giving them high grades, because I anticipated that with the other [civic world] jury members subtle judgements would get lost.*

This practice of strategic voting was noticed by Margareth, one of the other panel members, who was less experienced in these kinds of decision process:

> *I filled out the form in all honesty. But I saw that the urban planner was much more crafty. He used more extreme marks and thereby heavily influenced the score. I only saw through this in the third round and thought, 'If I really want to rule out a certain proposal because I think that will be a difficult person to work with, I also need to give extreme marks'.*

The results of the case studies confirm the trend that jury panels do not consist only of architectural professionals but often also include numerous stakeholders with different backgrounds. On a holistic level most of the decision makers in the cases appeared reasonably capable of making decisions, even if they felt insecure about their decision tasks. They just followed their intuition, which can be considered an appropriate strategy for ill-defined decisions in situations with a lack of information (Dane and Pratt, 2007). This was also mentioned by Bill, project leader of the democratic case, during a reflecting interview:

> *I think it is really clever of the award committee how they balanced all preferences. Although a pharmacist [members of the City Council are usually part time involved in politics and part time occupied by another profession] is not an urban planner, it did turn out well.*

Expertise is however also often limited to domains. A building project in a public context comprises several domains, which means a plethora of areas of expertise. Consequently it can be concluded that during architect selections decision makers need to be selected based on their competences, or being educated in performing their tasks. Especially in the critical feedback and learning curve of jury members and the issue of building trust, the concept of expert team (Salas, Burke and Stagl, 2004) has a lot to offer in the context of design tenders.

Justifying against different economies of worth

It was found that in order to make a decision, one does not only have to accurately analyse the complexity of the situation (judgement and computation), but also steer a course through the political reality of persuading others of the inspiration of an idea (negotiation) (Butler et al., 1993). In justifying a decision a decision maker is simultaneously confronted with the legal structure of the decision procedure and the iterative psychological organizational decision process.

The clash between different forms of value and different expectations about what constitutes a justified process was apparent not only to us during the analysis but was also articulated by participants in the cases, such as Dick, a participating architect in the democratic case:

> *Oh god, [there are] too many parties [involved]. It is all pseudo-democracy. You can see that the fear of being seen as undemocratic has increased, which has led to a kind of extremely transparent democratic processes [with the influence of many stakeholders].*

When comparing the two perspectives of the competition tradition and procurement law, it was found that they had very different thinking regarding the system rationale, the process approach and the decision outcome. According to procurement law the main aim of procuring a service is the allocation of a contract, which binds two legal entities for the establishment of a project. Architects are regarded as entrepreneurs rather than artists and it is the client who determines value maximization. The underlying reason for the frustration with current practice that many actors in our sample voiced, seems to stem not so much from a clash between the two selection methods per se, but from the different notions of worth and justification at the interface between the inspired, public opinion, industrial, market and civil worlds.

The dissatisfaction in the architecture community appeared to stem from a clash with the competition tradition they were accustomed to. Competitions constitute an established form of peer review consistent with the inspired world of Boltanski and Thévenot (2006). In the competition tradition the architect is considered to be an artist, to be judged anonymously by domain experts based on an artefact, the representation of the future building. The logic of competitions typically aims to avoid being swayed by public opinion and is not overly concerned with the rules and procedures of the civic world, as experts are expected to represent professional judgement. The competition tradition is based on the concept that submitting architects know enough about a competition if they know who is in the jury; knowing the jury members enables them to estimate whether the jury's artistic judgement will be aligned with their own artistic strivings. In the competition tradition, it constitutes no problem that worth evolves gradually in the sensemaking and comparing between the alternatives. The focus has traditionally been on the object with limited emphasis on the provision of a service and the allocation of a job. In our sample, this was apparent in the comments of Paul, an architect in the competition jury panel:

> *There are two ways of getting selected: based on the product, then you are taken hostage as architect but also protected by the promise of your design; or being selected as architect to make a building based on the trust of the client.*

The level of expertise and role of the decision maker play an important part in the justification of the decision because it makes it possible to separate from the context and scrutinize authoritative evidence, as also found by Gkeredakis et al. (2011) and Lamont, Mallard and Guetzkow (2006). The question, however, remains who should be considered an expert and how to deploy panel members for decision making. Furthermore, expertise is not always linked to power in public management.

Conclusion

This study addressed how clients negotiate different logics of justification and how they make sense of their task in selecting an architect for the design of a public building. Four cases of procuring architectural services as an example of high-stakes public strategic decisions were investigated and characterized as sensemaking dealing with different rationales. The analysis showed how these decisions involve the professional world, fostered by the inspired world and fame of architecture, and the civic world of procurement procedures and safeguarding common goods as addressed by Boltanski and Thévenot (2006).

Despite the differences between the four cases, the analysis reveals systematic similarities in the way decision makers and stakeholders try to establish a shared meaning about the party to award a contract to and gain the best of both worlds. Considering the mixture of aims, the involvement of multiple stakeholders and the political context, it is obvious that the relatively rational and static principles of procurement law clash with the professional world of architecture competitions. It was found that the assessment and justification frame of the decision developed during the process: in all cases architectural design played an important role as artefact in the decision process, and decision makers developed their own understanding about their needs and wishes for the future building in interaction with the design. However, this type of iterative process is in conflict with the legal requirement of making decision criteria transparent from the start. Jurors drew on arguments, perspectives and values as they fitted their intuitive judgement.

The implication of these findings is that decision makers in public organizations should draw on a combination of the competition tradition and the tender principles in their decision and justification processes. The actors in the cases showed that dealing with these decision logics is possible, but requires guts, creativity, perseverance and political sensitivity. In order to make decisions about any kind of partner selection in the public domain, some kind of system is needed. Yet, one could wonder if procurement – a large scale political system – provides the best system for appropriate and responsible decisions in the sensitive and complex public domain.

Compared to the artistic design tradition the legal procedures follow a rationalistic and quantitative approach, which is still seen as preferable from a societal perspective (Cabantous et al., 2010; Sinclair and Ashkanasy, 2005). A rationalistic decision motivation appears to fulfil the legal requirements of the civic world, but fails to include the process of collective sensemaking that client representatives undergo to reach their decision and that helps them to build support among the stakeholders in their inspired world.

The clash of different logics of justification also implies that different forms of expertise matter. While architects are trained to create an attractive and responsible built environment for the generations to come, representatives of public organizations need to spend public money wisely and negotiate the concerns of different stakeholders. The advance of procurement law has also resulted in expert power moving from architects to lawyers. It may be questionable whether the attention given to procedural justice actually helps to ensure the quality of the built environment within the budgetary constraints.

Implications

The implications of these findings relate to specific elements of the design of the selection procedure, either in case of a design competition or a design tender.

First of all, the assessment criteria should allow for addressing the characteristics of the architect as a person, the proposed design, as well as the firm that they represent. Although the competition tradition still tends to focus on the proposed design, the current position of the architect in the construction process is inclined to include much more responsibilities than visualizing the potential building (Bos-de Vos, Volker, and Wamelink, 2014). This should be realized by decision makers before the selection procedure is initiated and not considered as a result of the sensemaking process during the tender.

Second, the role of expertise should be acknowledged in the whole process. It is advisable to involve domain specific experts in the process of decision making because their specific knowledge on architecture, sustainability, urban planning, indoor climate or other issues related to design quality increases the chances of including these issues in the assessment criteria. Furthermore, they are better at controlling product emotions and using intuition than novices (Rosen et al., 2008). During a procedure the holistic judgement of an expert committee incorporates potentially conflicting judgements. Creating room and flexibility in the decision making process for discussion and negotiation among the decision makers will also improve decision quality. Unfortunately in the current procurement climate it is often hard for competition organizers to convince purchase officers and legal departments of this necessity.

Regarding justification of the decision it is advisable to motivate a decision and communicate this to the specific stakeholder groups, while using the media that belong to the different worlds of worth. Argumentation that fits the language of a particular group, tends to support the acceptance a particular decision. This not only requires administrators and project managers to adjust the tone of their message, but also challenges domain specific experts to explain their reasoning to others. The traditional jury report might thus not always be the only way of communicating the outcomes of a design competition.

Finally, ensuring that client ambitions are heard is something that should be taken into account during a design related selection process. If external experts are involved, they should be familiarized with the client's ambitions. Aligning the frames of reference of the actors during preparation of the tender process will increase the odds of reaching a decision at the end of the process that fits the ambitions and aims of the client organization. In this context it is advisable to organize a pilot session with all committee members to align each other's decision frames and mental models (Chupin, 2011). Conversely, including a personal explanation by the architect (by means of a presentation or dialogue) will improve the clients' understanding of the proposal that is offered. Although this might go against the tradition of blind jury assessments, one could think of selecting three to five of the best designs and having them presented in front of the deciding committee members of the client organization. Doesn't that sound like the best of both worlds?

References

Balogun, J., Pye, A., and Hodgkinson, G. P. (2008). Cognitively Skilled Organizational Decision Making: Making Sense of Deciding. In G. Hodgkinson and W. H. Starbuck (Eds), *The Oxford Handbook of Organizational Decision Making* (pp. 233–249). New York: Oxford University Press.

Beach, L. R., and Connolly, T. (2005). *The Psychology of Decision Making*. Thousand Oaks, CA: Sage.

Boltanski, L., and Thévenot, L. (2006). *On Justification: Economies of Worth* (C. Porter, trans.). Princeton: Princeton University Press.

Bos-de Vos, M., Volker, L., and Wamelink, H. (2014). Exploring new business strategies in architecture. Paper presented at the ARCOM 2014, Portsmouth.

Bresnen, M. (2010). Keeping it real? Constituting partnering through boundary objects. *Construction Management and Economics*, 28(6), pp. 615–628.

Butler, R. J., Davies, L., Pike, R., and Sharp, J. (1993). *Strategic Investment Decisions*. London: Routledge.

Cabantous, L., and Gond, J.-P. (2011). Rational decision making as performative praxis: explaining rationality's Éternel Retour. *Organization Science*, 22(3), pp. 573–586.

Cabantous, L., Gond, J.-P., and Johnson-Cramer, M. (2010). Decision theory as practice: crafting rationality in organizations. *Organization Studies*, 31(11), pp. 1531–1566.

Christianson, M. K., Farkas, M. T., Sutcliffe, K. M., and Weick, K. E. (2009). Learning through rare events: significant interruptions at the Baltimore and Ohio Railroad Museum. *Organization Science*, 20(5), pp. 846–860.

Chupin, J.-P. (2011). Judgement by design: towards a model for studying and improving the competition process in architecture and urban design. *Scandinavian Journal of Management*, 27(1), pp. 173–184.

Dane, E., and Pratt, M. (2007). Exploring intuition and its role in managerial decision making. *Academy of Management Review*, 32(1), pp. 33–54.

Flyvbjerg, B. (2004). Five Misunderstandings About Case-Study Research. In C. Seale, G. Gobo, J. F. Gubrium and D. Silverman (Eds), *Qualitative Research Practice* (pp. 420–434). London and Thousand Oaks, CA: Sage.

Gioia, D. A., and Chittipeddi, K. (1991). Sensemaking and sensegiving in strategic change initiation. *Strategic Management Journal*, 12(6), pp. 433–448.

Gioia, D. A., Corley, K. G., and Hamilton, A. L. (2013). Seeking qualitative rigor in inductive research: notes on the Gioia Methodology. *Organizational Research Methods*, 16(1), pp. 15–31.

Gkeredakis, E., Swan, J., Nicolini, D., and Scarbrough, H. (2011, 7–9 July). Rational decision making revisited: practices of deliberation in healthcare funding decisions. Paper presented at the 27th EGOS colloquium, Gothenburg.

Harrison, E. F. (1999). *The Managerial Decision-Making Process*. Boston, MA: Houghton Mifflin.

Hartmann, A., Davies, A., and Frederiksen, L. (2010). Learning to deliver service-enhanced public infrastructure: balancing contractual and relational capabilities. *Construction Management and Economics*, 28(11), pp. 1165–1175.

Hodgkinson, G., and Starbuck, W. H. (Eds) (2008). *The Oxford Handbook of Organizational Decision Making*. Oxford: Oxford University Press.

Hutton, R. J. B., and Klein, G. (1999). Expert decision making. *Systems Engineering*, 2(1), pp. 32–45.

Internal Market and Services Directorate General of the European Commission. (2011). *Commission Staff Working Paper: Evaluation Report – Impact and Effectiveness of EU Public Procurement Legislation (Vol. Part 1)*. Brussels: European Commission.

Jones, C., and Livne-Tarandach, R. (2008). Designing a frame: rhetorical strategies of architects. *Journal of Organizational Behavior*, 29(8), pp. 1075–1099.

Kazemian, R., and Rönn, M. (2009). Finnish architectural competitions: structure, criteria and judgement process. *Building Research & Information*, 37(2), pp. 176–186.

Kreiner, K. (2007). Constructing the client in architectural competition – an ethnographic study of revealed strategies. Paper presented at the EGOS 2007, Crete, Greece.

Kreiner, K., Jacobsen, P. H., and Jensen, D. T. (2011). Dialogues and the problems of knowing: reinventing the architectural competition. *Scandinavian Journal of Management*, 27(1), pp. 160–166.

Lamont, M., Mallard, G., and Guetzkow, J. (2006). Beyond blind faith: overcoming the obstacles to interdisciplinary evaluation. *Research Evaluation*, 15(1), pp. 43–55.

Langfeldt, L. (2001). The decision-making constraints and processes of grant peer review, and their effects on the review outcome. *Social Studies of Science*, 31(6), pp. 820–841.

Langley, A. (1999). Strategies for theorizing from process data. *The Academy of Management Review*, 24(4), pp. 691–710.

Langley, A., Smallman, C., Tsoukas, H., and Van de Ven, A. H. (2013). Process studies of change in organization and management: unveiling temporality, activity, and flow. *Academy of Management Journal*, 56(1), pp. 1–13.

Maitlis, S. (2005). The social processes of organizational sensemaking. *Academy of Management Journal*, 48(1), pp. 21–49.

Maitlis, S., and Sonenshein, S. (2010). Sensemaking in crisis and change: inspiration and insights from Weick (1988). *Journal of Management Studies*, 47(3), pp. 551–580.

Manzoni, B. (2014). Competing through architectural competitions: paradoxes and strategies. (PhD thesis), University College London.

March, J. G. (2006). Rationality, foolishness, and adaptive intelligence. *Strategic Management Journal*, 27, pp. 201–214.

Mieg, H. A. (2008). Professionalisation and professional identities of environmental experts: the case of Switzerland. *Environmental Sciences*, 5(1), pp. 41–51.

Miller, S. J., and Wilson, D. C. (2006). Perspectives on Organizational Decision-Making. In S. R. Clegg, C. Hardy, T. B. Lawrence and W. R. Nord (Eds), *The SAGE Handbook of Organization Studies* (2nd ed.). London: Sage.

Rosen, M., Salas, E., Lyons, R., and Fiore, S. M. (2008). Expertise and Naturalistic Decision Making: Mechanisms of Effective Decision Making. In G. Hodgkinson and W. H. Starbuck (Eds), *The Oxford Handbook of Organizational Decision Making* (pp. 211–230). New York: Oxford University Press.

Salas, E., Burke, C. S., and Stagl, K. C. (2004). Developing Teams and Team Leaders: Strategies and Principles. In D. Day, S. J. Zaccaro and S. M. Halpin (Eds), *Leader Development for*

Transforming Organizations: Growing Leaders for Tomorrow (pp. 325–355). Mahwah, NJ: Lawrence Erlbaum Associates.

Simon, H. A. (1997). *Administrative Behavior: A Study of Decision-Making Processes in Administrative Organizations* (4th ed.). New York: MacMillan.

Sinclair, M., and Ashkanasy, N. M. (2005). Intuition: myth or a decision-making tool? *Management Learning*, 36(3), pp. 353–370.

Strong, J. (1996). *Winning by Design – Architectural Competitions*. Oxford: Butterworth Architecture.

Thompson, D. F. (1980). Moral responsibility of public officials: the problem of many hands. *The American Political Science Review*, 74, pp. 905–916.

Tversky, A., and Kahneman, D. (1981). The framing of decisions and the psychology of choice. *Science*, 211, pp. 1124–1131.

Volker, L. (2010). *Deciding about Design Quality – Value Judgements and Decision Making in the Selection of Architects by Public Clients under European Tendering Regulations*. Leiden: Sidestone Press/Delft University of Technology.

Volker, L., Lauche, K., Heintz, J. L., and de Jonge, H. (2008). Deciding about design quality: design perception during a European tendering procedure. *Design Studies*, 29(4), pp. 387–409.

Weick, K. E. (1995). *Sensemaking in Organizations*. Thousand Oaks, CA: Sage.

White, J. T. (2014). Design by competition and the potential for public participation: assessing an urban design competition on Toronto's waterfront. *Journal of Urban Design*, 19(4), pp. 541–564.

4 Design in conversation

An interview with Malcolm Reading

Chapters 1 to 3 have revealed that competitions are design processes that need contextually specific, rather than strictly formalized ways of organization and execution. But what does that mean and how would this insight into the need for informality encourage potential clients to decide that a competition is the right procedure to realize a particular building task? In the concluding interview for Part I we open up on these questions by focusing on the everyday practice of competition organization and ask how an organizer responds to and navigates various global and local building cultures. Our interview partner for this closing chapter of Part I is Malcolm Reading, a professional competition organizer with experience in organizing competitions for public awarding authorities, as well as for private clients, who shows awareness of the in-situ character of competitions and accordingly shapes and adapts his organizational work to the specific needs of particular local situations and circumstances.

An interview with Malcolm Reading, Chairman of Malcolm Reading Consultants, London, UK

Malcolm Reading Consultants (MRC) provides pre-project services, briefing and strategic advice on capital projects in the UK and worldwide. MRC's international portfolio includes competitions for Queen Elizabeth Olympic Park's Culture & Education Quarter; the Mumbai City Museum; Guggenheim Helsinki; Glasgow School of Art; and the Library of Foreign Literature in Moscow. Since its foundation in 1996, MRC has worked for diverse organizations including the Foreign and Commonwealth Office; New College, Oxford; Qatari Diar; the Cadogan Estate; Marlborough College; the Natural History Museum; White Cube; the National Media Museum; NATO; the Duchy of Lancaster; and the Victoria and Albert Museum. Over the last decade, MRC's competitions have brought emerging designers to world attention, including Thomas Heatherwick, Amanda Levete and Steven Holl.

Malcolm Reading, MRC's Founder and Chairman, trained as an architect and moved from designing to commissioning designers with his appointment as Director of Design at the British Council in the early nineties. Today, MRC includes eight staff members with varied backgrounds in architecture, urban design, graphic design, web design and journalism. Additionally, several staff members have both commercial and public sector experience.

Design in conversation 79

Malcolm Reading (MR) was interviewed by Jan Silberberger (JS).

JS: Tell us about the origins of MRC.

MR: When I came out of my former work at the British Council, there was a moment when I realized that a lot of clients, who don't build very often, have no idea how to commission an architect. And there were a lot of buildings in the arts and culture field with quite complex users. So the idea was that I might help such clients find architects – and from that grew my work in competitions.

JS: It seems that most competitions you are currently organizing are located in the USA.

MR: They are all around. As of today we are about to launch a competition in Qatar for a big new museum. Then one of my colleagues is going to Washington DC for a master plan for a university that teaches deaf people, but we are also about to launch a competition for a bridge in Cornwall. Competitions tend to go where the project is – that's not really something we can influence.

JS: How do you acquire new clients? Or is it the case that clients approach you?

MR: The contacts are more and more coming to us. I think competition organizers in Central Europe have an easier time because having competitions is mandatory there, but overseas in America they don't have a developed public competition culture. We had been recommended there through architects who knew our work. The way we often get introduced to clients is through well-known architects.

JS: On your website you address the problem of different procurement regulations in the various countries where you operate. Would you say that it is easier to organize a competition in the USA than in Europe?

MR: You have regulations everywhere. In the USA one has to be aware of the fact that the different federal states have different approaches, so it's not as easy as in Europe. But also in Europe, the ways we manage European Union regulations in the UK are so different to how they are interpreted in Germany or France. There is one set of European Union regulations, but everybody reads them in very different ways. But we also have experience in places such as India and that's a totally different story. There, the competition process is difficult to overlay on their standard procurement procedures. There, if you provide a correct technical proposal, you get through to the next stage, where the one who offers the lowest fee is selected. However, we argue that in a competition there is crossover analysis, where you are always balancing quality with value and costs. Organizing competitions in such contexts is quite difficult. Equally some clients are afraid of European Union regulations because they think they are going to force them down a particular route, but they're not really. If you lay out what you are trying to achieve, if you are frank about how things are going to be assessed, and you stick to that – then this is the ethos of the EU process.

JS: One of the core sources of conflict in design competitions is the 'honourable mention' that jury boards use to select an entry which produces a convincing solution to the problem, thereby redefining the scope given in the brief. This practice, which is essential to design work, constitutes a source of conflict regarding the principles of public procurement regulation – transparency, equal information and non-discrimination.

> MR: I think there is a fundamental difference to the way we prepare and use briefs. The way that we approach a competition is we try to find the right architect for the right client. It is a process that's a bit like headhunting. You use design as a way to have a conversation, a dialogue. But you are not judged solely on the design achieving certain parameters or whether it's got the right number of toilets in it, or if reception is in the right place, because that will follow as a good design progresses to maturity. Our process is more reliant on a series of skills, which the competition unveils. It's not about whether the brief has been met in the particular case. However, we know that this approach is not universally accepted – an example is the Guggenheim competition in Helsinki. The initial reaction from the Finnish Society of Architects to our proposal was an objection that the jury would meet the teams. And we were adamant that they should. In the end we agreed a compromise where the teams could be announced and their anonymous designs are known but you can't put the two together until the final moment of selection.

JS: But when conversation and dialogue are so central, are you able to treat a young, inexperienced office the same way as a big, established architecture firm?

> MR: People often use that argument. I think that is not respecting the jury. Say you have a young team and you are the client and really believe in their work, then you find somebody to help them out, be that an engineer or an executive architect. In my experience the people we have on juries are highly intelligent and very engaged with the client. And when we have a problem like 'Your practice is only five strong', or 'You are in Switzerland and we are in London' or something similar, you have a conversation and you discuss what has to be done about that.

JS: What is your role in such discussions?

> MR: I am part of the jury. I chair the jury often, but in the sense of chairing the administration of the process to make sure everybody sticks to the time schedule and everybody has a chance to raise questions. I assist the chair of the jury and I ask questions. Ultimately, the aim should be for a jury to agree by consensus. We've never had a vote to majority. You've really lost if you have to vote.

JS: How do you work with the jurors?

> MR: Three or four weeks before the jury sessions, jurors will be sent a letter and a written understanding of what we expect from jury work. They get the brief, and then about two weeks before the jury board meeting, they get the submissions together with a detailed but objective report from various specialists – we call this a peer-review report. We find the jurors get more and more interested and this engagement helps things go well when we finally all meet. Important is the number of jurors. In the Guggenheim competition, we had twelve jurors and that is the absolute limit I would say. Seven is my perfect number. You can have a conversation around the table – it's like dinner. But if you get more, any more than nine,

interactions become more difficult to manage. But juries work in different ways really. Some juries like to remain formally around the table, and they don't like to get up. And then some juries are much more interactive and move boards around.

JS: You have mentioned how you gain the interest of jurors. How do you interest architecture offices in the competitions that you organize?

MR: We have a database of 6,000 architects and we particularly contact young architects and engage with them. We are now using a form system to get more young architects to register with us and to tell us what they do. So these 6,000 architects all get a personal flyer about the competitions we organize. And there are five or six portals around the world like Bustler, Death by Architecture and Architecture Daily, which are the biggest referrals to the competition site.

JS: What do architects have to hand in on the first stage?

MR: A lot of our competitions in the first stage do not ask for designs. We ask questions about why a team's experience is relevant to the brief and what they think about the brief. It's not a design, but a communication response, written and illustrated, twenty to twenty-five pages. That is quite helpful, it weeds out people who just print out their marketing literature and send it. We've already got their focus on the brief, so you are in dialogue with people who actually have gone to see the site and have read what you want to achieve. I'd say it's a less wasteful and much more productive way than trying to pick from 100. Plus how can you write an adequate brief until you've been through that process? We always have a stage one competition brief to get special interest, and then the six selected teams get a different brief, a better developed one.

JS: What is significant in regards to your competition briefs is that they completely do without the usual detailed specification lists.

MR: You are talking to people who communicate visually. There is little value in writing out series of Excel sheets. You can't avoid all information – the size of the building, the size of the spaces, the relationships have to be listed – but then we take that list and try to turn it into something which communicates better. The trouble with an Excel sheet is that it looks like information that cannot be changed. As soon as you translate this information into a visualization, working with colours, there starts to be a certain dynamic so that the reader sees scope for intervention. It's a method of communication.

JS: But what about very complicated buildings? Does this work, too?

MR: A hospital is a situation where you can't have a loose brief. If you have a special care unit that has to work in a particular way because there is a process that it has to reflect. The competition for a complex building really needs to be focused on finding the right architect and not the right design, because the design is going to evolve; so you really don't want to freeze it at a certain point where it is very young, adolescent. Generally, I would say it is a surprise if a competition scheme gets built exactly as it won the competition. Life is not like that. And so

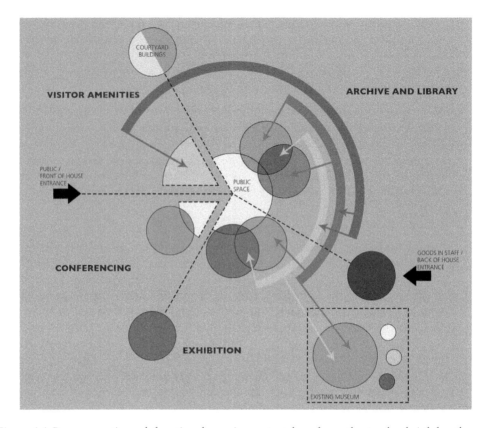

Figure 4.1 Programmatic and functional requirements taken from the tender brief for the Mumbai City Museum North Wing.
Source: Malcolm Reading Consultants.

I think the key is, the concept, the core of the concept remains but the execution may be quite different. And the team should remain. We have one building that hasn't been built from all of our twenty years in competition. There must be something good about our process.

JS: Can you provide an example of a project that significantly changed after the competition?

MR: We had a competition in Greenwich, in the National Maritime Museum, which C. F. Møller won, on which we subsequently remained as adviser to the client. That did change significantly because it is a World Heritage Site and there was quite an involvement with changing parts of the building for heritage impact. So the overall concept remained but the visualization externally changed. We've had a number of cases where projects got much better after the competition.

JS: This means that you are following the project in the building phase?

> MR: I would say that in 50 per cent of the cases we are asked to carry on as an advisor. Sometimes that is up to submission of planning, so that's about a year after the competition, and in many cases it's all the way to the end of the project, just helping the client to really keep their objectives.

JS: It is not uncommon that winning entries change to the extent that they lose some of the qualities that the jury picked them for.

> MR: That's why all of our competitions have what we call 'after-care'. I feel the absolute worst moment is when you have a client, maybe a client who's never designed before, and they spend a lot of money on a competition, they have the jury, then five o'clock comes on the jury day, everybody shakes hands – and walks out the door! The client is left there with an architect, and they really don't know what to do next: there is nobody else to help. Of course I am exaggerating to make a point but my theory is, during the first six to eight weeks, the client and the architect need some help to get to know each other. Some architects are good at it, but a lot aren't. And if a client really has no experience of the design process then it seems to me that you can have quite a lot of value at that point. It is not a project management issue – it's a human issue.

JS: In the end learning to design is also fundamentally about learning to judge design.

> MR: It is nice to have observers as part of the jury, if it's a small enough number. I remember we had five or six people at the Guggenheim. They were all design students and they were just amazed and fascinated to hear how the process unfolded, the way each project was dissected and analysed. In my view this is in sharp contrast with their experience with the system of architectural education that relies very much on crits: you know, you've done three months where you haven't slept at all for the last two months, you pin your design up, then somebody dressed in high-fashion black enters the room and just sort of beats you. It's a very aggressive, unpleasant way of learning and inevitably this gets carried out to the working world. However, I would be careful with opening up completely. I would say that to have a public jury would ruin the way that a jury works. I think a competition works not only because it provides for an ideal space of judgement, but because it builds also on the agreement that the jury is competent to select the right project and that the way the decision has been taken shall not be retraced.

Part II
Inside the competition

5 The progressive differentiation of judgement criteria

Jan Silberberger and Ignaz Strebel

Introduction

Briefs of architectural competitions are often structured in a rather similar manner. Often they start with providing an initial position and description of the objective. Then they provide information about the formal regulations, the stakeholders, the procedure and the deadlines. Next, there is an in-depth description of the task, usually consisting of a more detailed depiction of the project and the perimeter. This is followed by general terms and conditions such as the project's urban-design and planning significance, building and planning law specifications, traffic development and subsoil. Then there are sections concerned with programmatic specifications and givens such as spatial specifications, technical requirements, target costs and energy efficiency. After that there is a chapter on the criteria according to which the entries are to be judged as well as a list of the jury board. Finally, the brief lists the documents to be submitted.

What is common to competition briefs in solution-orientated competitions is that the sections regarding for instance spatial specifications are extensive and as specific to the task as possible, while those sections dealing with judgement criteria are most of the time rather short and (often) could not be more nonspecific. To support this argument we will show below two excerpts from the brief for 'The Basel Kunstmuseum, Burghof Extension' competition. This anonymous project competition (in a selective procedure) took place from 2009 to 2010 and concerned a major extension to the famous museum for fine arts in the city of Basel. The first excerpt shows the first page of a two-page overview of the spatial specifications (Figure 5.1). The second features the whole chapter on judgement criteria (Figure 5.2).

Even a rough glance at these two excerpts will reveal that while participants are precisely instructed on the task, there is not even an attempt to define judgement criteria more accurately. One may go so far as to say that the list of judgement criteria featured in the second excerpt – urban design and planning (integration into the existing built environment), architecture, functionality, operating feasibility, compliance with the budget, cost-efficient maintenance and operation as well as ecological sustainability – could be applied to every project competition around the globe. It is comprehensible that due to this nonspecific, across-the-board definition of judgement criteria architects wondering whether they should enter a competition primarily base their decision on the composition of the board of jurors or on their guesses in regard to jurors' architectural preferences.

Use, Area	Function	Note: type, number, and quality of room	room dimensions m x m *italics: reference dimension*	Surface NGF (EFS) m2	Clearance m	Room, Division	Room relation, operational concept UF / GF / BF *also see additional drawings, processes*	Location UF / GF / BF *as far as com-pelling*	Airconditioning zone *see p. 35*	Safety zone	Light Daylight DL Artificial light AL Skylight SL
°1	**Exhibition**			**3 200**							
1.1	Special exhibition	Basic module	Types: small 4x medium 2x large 1x	*7 x 9 = 63 10 x 15 = 150 13 x 19 = 247*	800	6.0	no mobile partition walls, no cul-de-sacs	Modules combinable, on a *single* floor	UF	3+	2	SL
1.2	Special exhibition	Extension module	Types: small 4x medium 1x	*7 x 9 = 63 10 x 15 = 150*	400	6.0		expands basic module or presen-tation of collection				
1.3	Presentation of collection	Presentation of the museum collection	Types: small 6x medium 4x large 4x	*7 x 9 = 63 10 x 15 = 150 13 x 19 = 247*	2 000	>5.0 5.5-6.0			GF, UF			DL, AL, SL
°2	**Development with foyer**			**1 470**							
2.1	Foyer	Events, private views	for ~1000 people, with ~300 seats, entire building accessible to the handicapped		600	≥5.5	A *single* surface, if possible without supports, inc. 20m2 information desk, shop, special exhibition	possibly combinable with connecting wing		3	2	Foyer: DL
2.2	Connecting wing	generously proportioned connection to the main building		minimum of 15 m wide	450	≥3.8		possibly as an extension to the foyer	Recom-mendati on BF	3	1	AL
2.3	Development	to floors and galleries > pre-zone	Generous pre-zones		depends on					3		
		Goods lift	per floor 45m2 Assumption: 6 - max. 7 floors	Cabin app. 8 x 5 +well	315	3.8		Delivery				
		Visitors' lift	~3 lifts of 5m2, each for 10 people, Assumption : 6 - max. 7 floors	Cabin app. 1.5 x 2 +well	105							
°3	**Delivery and parking spaces**			**505**							
3.1	Delivery	Lock	Hall with lift table, level to the floor, lowerable, for articulated lorries 18.75m, in front of or side by side of each other	10 x 25	250	≥4.5		Process, see schematic diagram, well possible as foyer extension	EG	3	5	AL
		Quarantine zone	Quarantine room, interim storage, transport boxes	4 x 10	40			Expansion foyer			6	AL
			Work rooms, art handling - opening - logging - presentation, photo studio		100						4	DL
		Intermediate store of boxes	empty boxes, box mass because of lorries		50						4	DL
3.2	Bicycle stands	Staff	for 40 people, roofed over					close to entrance lodge		−		
3.3	Parking lots	for cars	2, always accessible		25			as closely together as possible				
3.4	Short-term parking lots	for small lorries	2-3 desired		40							
°4	**Depot**			**1 540**							
4.1	Depot	Paintings, large sizes, sculptures	~2/3 with hanging system	15m width	800	4.5		Proposed 2nd BF	UG	3a-g	6	KL
		Graphics, large sizes	hanging system	15m width	100							
		Grafiken	app. 3, height >4.5m, with filing cabinets	12 x 15 floor space	300	≥4.5	with mezzanine					
		Photographs, documents, videos, films	app. 3, height 4.5m, with filing cabinets or 4 Kompaktus "cold-store cells" for videos, 16mm b/w and colour, CDs	4 "cold-store cells" 4m2	300							
		Frames			40							

Figure 5.1 The first page of a two-page overview of the spatial specifications.
Source: 'The Basel Kunstmuseum, Burghof Extension' competition programme, p. 22. Courtesy of Beatrice Bayer Architekten (Basel) and Bau- und Verkehrsdepartment des Kantons Basel-Stadt, Hochbau- und Planungsamt.

Evaluation and Assessment Criteria

The projects are evaluated and assessed according to the criteria listed below.

- Urban design and planning, integration into the building history and context
- Architecture
- Functionality of the museum operation, meeting the Spatial Specifications
- Operating feasibility
  Compliance with the budget
  Cost-efficient maintenance and operation
- Ecological sustainability
  Energy and ecological materiality according to Minergie-P and Minergie-Eco specifications

The sequence of criteria does not equal a valuation.
The jury will make an overall assessment and evaluation on the basis of the above-mentioned criteria.

Figure 5.2 Evaluation and assessment criteria for the 'Basel Kunstmuseum, Burghof Extension' competition.
Source: 'The Basel Kunstmuseum, Burghof Extension' competition programme, p. 33. Courtesy of Beatrice Bayer Architekten (Basel) and Bau- und Verkehrsdepartment des Kantons Basel-Stadt, Hochbau- und Planungsamt.

This, however, is not the point to be elaborated in this chapter. Instead, we will focus on the practices inside jury boards and aim at showing how a jury's perspective on the problem at hand – that is, its perception of quality as regards urban design, architecture and functionality – evolves throughout the jury deliberations. Drawing on data gained by means of a participatory observation of the jury boards of four architectural competitions in Switzerland (see Van Wezemael, Silberberger and Paisiou, 2011), we will demonstrate that this process of differentiation is highly dependent on certain thought-provoking entries and – most importantly – does not necessarily proceed straight-line, but may exhibit significant shifts and leaps (Chupin, 2011; Kreiner et al., 2011; Van Wezemael et al., 2011). In order to make that point, we will present four vignettes, that is, short, concentrated, self-contained descriptions of situations, which we (re-)constructed on the basis of our field-notes (Söderström, 2000). By means of these four vignettes we will explore and theorize the differentiation of a jury's perception of quality as an interaction or a co-production between the board of jurors and distinct properties of certain competition entries as displayed on architectural plans or featured on architectural models. Introducing and applying the concept of sensemaking (Weick, 1995), our analysis closely relates to Kreiner's work on architectural judgement as sensemaking (Kreiner et al., 2011; Kreiner, 2010).

The case – 'The Basel Kunstmuseum, Burghof Extension' competition

On 21 February 2009, the City of Basel officially advertised the 'Basel Kunstmuseum, Burghof Extension' project competition. The task posed concerned the new construction of a major extension (with a net internal area of approximately 7500 square metres) for the existing museum, which displays Basel's most important collection of fine arts. Figure 5.3 shows the Basel Kunstmuseum, which was built by architects Rudolf Christ and Paul Bonatz (the building was completed in 1936), on the left and the Burghof lot – the site for the extension – on the right. The targeted vision was to:

> enhance the Kunstmuseum into a venue radiating tradition while remaining a dynamic and open institution of international renown with strong local roots, too. Not only the collection but also travelling exhibits and special shows of the museum are constantly on a world-class level that is to be consolidated and expanded.
> ('The Basel Kunstmuseum, Burghof Extension' competition brief, page 3)

The competition was carried out in a restricted procedure and 134 architecture offices handed in their applications before 3 April 2009 with 49 out of these 134 offices being based in Switzerland (see Van Wezemael and Silberberger, 2015). On 27 April, the jury board selected twenty-four offices to provide solutions to the task posed, with fourteen out of these twenty-four coming from Switzerland. On 18 May these selected teams were invited to a site inspection and the project competition officially started. Out of the twenty-four selected offices (four of them 'newcomers') twenty-three handed in the required documents by 18 September 2009. The jury assessment sessions took place on three days: on 5, 6 and 20 November. The jury's final report was published on 7 December 2009 and all competition entries were displayed to the public in an exhibition from 8–20 December.

An important aspect as regards the task posed was that the extension was:

> to be accessed by way of the main building. Visitors shall enter the Kunstmuseum by way of the large interior courtyard and reach central services such as the pay desk, wardrobe, bistro and bookshop serving all parts of the museum both in the existing and the future building.

Furthermore, the passage between the main building and the extension was 'envisioned as a generously proportioned structural element, which has to meet a large range of requirements. Instead of a simple passageway, visitors are to cross through a space housing artistic presentations, for instance, one presenting new media' ('The Basel Kunstmuseum, Burghof Extension' competition brief, page 7; compare Figure 5.3 and see also Figure 5.4). Crucial for our analysis is the fact that the brief recommends placing the connecting wing on the first basement floor and states that '[a]n aboveground connection is possible but not recommended' (competition brief, page 28; see Figure 5.4).

When taking these statements into consideration, participating architecture offices, and generally speaking everyone who had read the brief, had concluded that the board of jurors, having approved its contents, shared the following perspective: an aboveground connecting wing was problematic due to issues of urban design as well as issues of functionality. Hence, one could argue that the rather nonspecific criteria 'Urban design and planning, integration into the building history and context', 'Architecture' as well as 'Functionality of the museum operation, meeting the spatial specifications' (see Figure 5.2) were in fact determined to a certain extent in the sense that the brief provided a vague definition of quality in regard to these three criteria. In what follows, we will show that this definition will change dramatically throughout jury deliberations as judgement criteria evolve by means of the interplay of jurors and competition entries.

Differentiation of judgement criteria 91

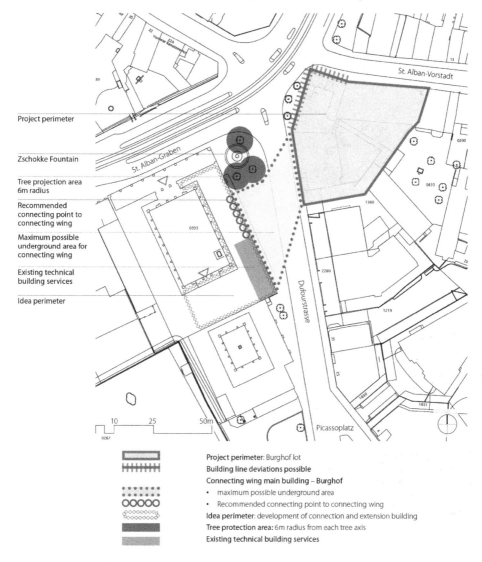

Figure 5.3 Diagram showing the project perimeter.
Source: 'The Basel Kunstmuseum, Burghof Extension' competition programme, p. 8. Courtesy of Beatrice Bayer Architekten (Basel) and Bau- und Verkehrsdepartment des Kantons Basel-Stadt, Hochbau- und Planungsamt.

The four vignettes

By means of the four following vignettes we will follow a controversial competition entry – 'Projekt 2: Neunhundertdreiundvierzig' (Figure 5.5; 'Neunhundertdreiundvierzig', the number 943 spelled out in German, is the name chosen by the competitors, 'Projekt 2' is an ascription by the organizers; we will hereafter refer to this entry as *Entry#2*) – through three days of jury assessment sessions. As can be seen in Figure 5.5, which represents plans one and three of the

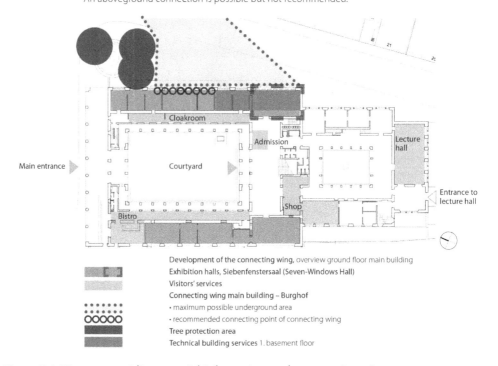

Figure 5.4 Diagram providing essential information on the connecting wing.
Source: 'The Basel Kunstmuseum, Burghof Extension' competition programme, p. 26. Courtesy of Beatrice Bayer Architekten (Basel) and Bau- und Verkehrsdepartment des Kantons Basel-Stadt, Hochbau- und Planungsamt.

six A0-plans Basel based architecture office Diener & Diener Architekten submitted, *Entry#2* suggests an aboveground connection between the existing main building and the extension (besides *Entry#2*, of the twenty-three submissions four other entries proposed an aboveground connection). This two-storey connecting wing is covered with wood, which makes *Entry#2* an exception; it is the only entry using wood as the material of choice for a part of the façade.

Vignettes 1, 3 and 4 refer to situations in which the board of jurors gather in front of the six A0-plans (and the architectural model, which is in a scale of 1:500) representing *Entry#2*; Vignette 2 refers to a situation in which all jury members sit at a round table having the brief and the competition entry screening report in front of them. To simplify matters, we have designated each jury member an alphabetical character in order of appearance.

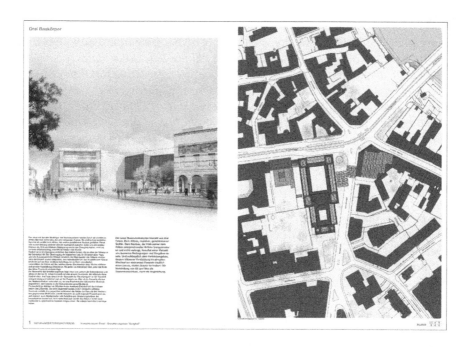

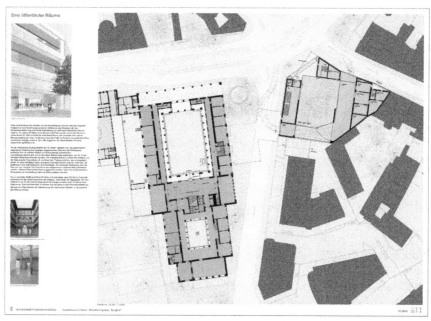

Figure 5.5 'Projekt 2: Neunhundertdreiundvierzig', *Plan#1* and *Plan#3* (original plans presented in the competition measured 118.8 cm × 84.1 cm).
Source: 'The Basel Kunstmuseum, Burghof Extension' competition programme, jury report, pp. 53–54. Courtesy of Beatrice Bayer Architekten (Basel) and Bau- und Verkehrsdepartment des Kantons Basel-Stadt, Hochbau- und Planungsamt. Architect: Diener & Diener Architekten, Basel, Switzerland.

Vignette 1

November 5th 2009, first day, first round of the jury assessment sessions, ca. 10.30 am. After the first entry has been eliminated rather quickly and almost without a dissenting vote, this seems to be the case, too, for Entry#2. Several jurors object to the façade Entry#2 proposes. Most jurors do so in a rather rigorous manner. Juror A, pointing to the left hand side of Plan#1, states: 'This façade looks provincial!' Juror B adds: 'This wooden box is an aberration! It disturbs the existing urban structure.' Another three jurors stress that they consider the aboveground connecting wing problematic. First, due to aspects of urban design – 'an aboveground connection interferes with the street as a defined space' (Juror C). And second, since it 'destroys the sequence of rooms of the Bonatz building' (Juror D, pointing to Plan#3), which, as Juror A adds, 'belongs to the best on the planet!' Only one juror – Juror E – (half-heartedly) tries to defend Entry#2, mentioning that he appreciates the 'provisional character of the wooden box' and that he sees a 'certain' quality in connecting the two buildings aboveground. Yet, his argumentation comes to nothing. Without much further discussion the jury eliminates Entry#2 after not even ten minutes of discussion.

Vignette 2

November 6th 2009, second day of the jury assessment sessions, review session, ca. 8.45 am. Before starting with the second round, the board of jurors reviews its work of the previous day. While discussing the decisions taken, most jury members leaf through the competition entry screening report that provides an overview of all entries. Suddenly, Juror F makes the motion to reassess Entry#2: 'It is the only entry that convincingly suggests an aboveground connection. Maybe we just need this entry in order to see more clearly what we are looking for. But maybe we should really reconsider this proposal. That wooden box is like a third building that connects the existing museum and the extension.' [In fact at this point in time, one other entry proposing an aboveground connecting wing – Entry#16 – remains in the competition.] Juror E, who tried to argue in favour of Entry#2 on the first day, immediately supports this motion. After only a very brief discussion the jury decides to take Entry#2 back into the competition.

Vignette 3

November 6th 2009, second day of the jury assessment sessions, second round, ca. 4.45 pm. [Entry#2 is, due to the fact that the jury reversed order for its second round, the last entry to be judged on this day.] The jury president opens the discussion. He points out that Entry#2 is the only entry left that suggests an aboveground connecting wing. [The jury dismissed Entry#16 earlier in the second round.] Hereupon several jurors mark the aboveground connection as a 'disruption' (Juror A) and as a 'no-go' (Juror C) repeating the arguments put forward in the first round. Then Juror F, who made the motion in the morning, speaks. Pointing to Plan#3, he identifies the aboveground connection as a pivotal achievement: 'Entry#2 connects gallery spaces with gallery spaces via gallery spaces. That is brilliant!' After a short pause, he adds: 'This is probably the most unpretentious entry. Everything it suggests is gallery space.' Juror E again supports him immediately: 'This is visible from the outside. The wooden interface building is probably rather consequent.' Soon more jurors share this perception. The jury divides into two irreconcilable groups of almost the same size. The debate becomes heated and/but the incompatibility of the two perspectives remains. Finally, the jury president asks the jury members to vote by a show of hands for or against taking Entry#2 to the next round. The result of that voting is six against six. The jury president, who did not take part in the voting, eventually decides to take Entry#2 to round three.

Vignette 4

November 20th 2009, third day of the jury assessment sessions, ca. 1 pm, in the middle of the third round. While Entry#2 had split the jury into two equally sized camps in the jury's second round, this balance now gets destabilized. Almost all jurors recognize the advantages of aboveground gallery spaces functioning as a connecting wing. And nearly the whole contra-alliance can be convinced that an aboveground connection is not to be regarded as a severe problem: neither in regard to the existing urban structure nor to the existing sequence of rooms – a prominent exception being Juror C. Pointing to the architectural model, he keeps insisting that 'an aboveground connecting wing is not only an encroachment on the existing Bonatz building itself but also a severe disturbance of the surrounding quarter'. Yet, Juror C stands rather alone now. His argumentation does not meet much response. The jury quasi-unanimously decides in favour of Entry#2. Eventually, Juror F mentions: 'If you look at it in retrospect, you have to say that the brief should have been different regarding the connection. That underground approach looks a bit drab now.'

Sensemaking as a collaboration of humans and artefacts

Before analysing Vignettes 1–4, we will provide a brief summary of Weickian theories of sensemaking. In doing so, we will touch on the concept of 'reflection-in-action' as developed by Schön (1983) in order to point to the opportunities both offer in regard to reframing Weick's arguments by strengthening the collaborative relation of human and non-human actors, which in architectural competitions is pivotal.

Sensemaking, 'the process through which individuals work to understand novel, unexpected, or confusing events' (Maitlis and Christianson, 2014, p. 57) on an organizational level is the main theme in Weick's multifaceted work. Weick's sensemaking started as a supporting concept for crisis management (Weick, 1988). However, in recent decades, a large body of literature has accumulated in regard to sensemaking 'in more conventional contexts' (Maitlis and Christianson, 2014, p. 61). Thus, its application can be found in diverse organizational settings, ranging from planned change intervention and innovation to learning and knowledge-creation and more. For example, in their ethnographic study of efforts of strategic change in a public university, Gioia and Chittipeddi (1991) have described processes in terms of 'sensemaking and sensegiving' (p. 444), meaning that some members of the collective first need to make sense of relevant 'events, threats, opportunities' before, in a second step, they may propagate their point of view to other members (p. 444). In a similar vein, Kreiner et al. (2011) elaborate on the pivotal role that retrospective sensemaking plays in decision making processes of the jury boards of architectural competitions. Kreiner (2010) also – by means of an ethnographic study of architects dealing with competition briefs – elaborates on sensemaking as an integral part of designing a competition entry.

Common to Kreiner's and similar studies in other fields (see, for example, Balogun and Johnson, 2005; Gioia and Thomas, 1996) is a perspective on sensemaking as a process that comprises noticing, finding and 'bracketing cues in the environment, creating inter-subjective meaning through cycles of interpretation and action and thereby enacting a more ordered environment from which further cues can be drawn' (Maitlis and Christianson, 2014, p. 67). As is the case with sensemaking under pressure, sensemaking in slow-paced non-catastrophic environments is triggered by some novel event or experience that interrupts business-as-usual or the 'ongoing flow' (Weick, 1995, p. 100) of how things proceed.

To this we want to add a material dimension, as the handling of plans, documents and visualizations in architectural competitons is a crucial dimension of sensemaking. As Orlikowski and Scott (2008) have pointed out, there is still little scholarship on the roles things or artefacts may play in sensemaking processes. Useful for our purposes are Rafaeli and Vilnai-Yavetz (2004a; 2004b), who introduced and elaborated on a multidimensional model for artefacts, that is, they proposed analysing an artefact according to its instrumentality, meaning 'the extent to which the artefact contributes to performance or to promoting goals'; to its aesthetics, meaning 'the sensory experience it elicits and the extent to which this experience fits individual goals and spirit'; and its symbolism, 'the meanings or associations it elicits' (2004a, p. 94). Furthermore, Stigliani and Ravasi (2012) with their thorough investigation of material practices in product design provided a detailed analysis of the supportive role of (visualizing) material in collective (prospective) sensemaking processes. Yet, they argue from the point of view of cognitive psychology – here material artefacts can be seen as 'extensions of [a human's] cognitive resources' (Clark and Chalmers, 2010, p. 32) that become 'interactive tools' (Stigliani and Ravasi, 2012, p. 1253): materializations of thoughts and ideas into sketches, press clippings or PowerPoint slides that facilitate present and future availability (for others) and exchange, thereby enhancing individuals' and groups' ability 'to process mental content' (Stigliani and Ravasi, 2012).

In line with this scholarship, Schön's (1983) account of designers in 'conversation with the materials of [their] design[s]', with these materials 'continually talking back to [them], causing [them] to apprehend unanticipated problems and potentials' (p. 101) provides an important starting point for our study. Weick himself (1995, p. 9) acknowledges Schön's (1983, p. 40) notion of sensemaking that is needed to 'convert a problematic situation to a problem' that then may be solvable. An 'individual is trying to deal' with 'some puzzling or troubling or interesting phenomenon' and while trying 'to make sense of it he also reflects on the understandings which he surfaces, criticizes, restructures and embodies in further action' (Schön, 1983, p. 50). This is what Schön calls 'reflection-in-action'. Yanow and Tsoukas (2009) have pointed out that such processes of reflection are always 'embedded in practice activities'. Knowledge, as they point out, 'is acquired through active engagement in and with the practice world, not through thought alone' (p. 1347). This connotes a perspective on sensemaking that takes into account that the making, 'whether of things or of theories, is always [...] a matter of attending to the material conditions of its activity' (Harrison, 1978, p. 209).

In what follows, we will analyse architectural judgement as a collective achievement, that is, as a co-production of sense involving – in our case – a variety of jurors and an active artefact (most prominently in the vignettes on *Entry#2*, which consists of six plans mounted on a movable wall and an architectural model positioned on a nearby table). As we go along and discuss the vignettes presented above we intend to clarify that sensemaking and generation of knowledge is not only an affair of agency between humans and non-humans in immediate judgement situations, but that judgement procedures and involved agents move back and forth during the building process.

Jury work as collective sensemaking

From a Weickian point of view sensemaking starts with a crisis or a perceived instability and therefore an apparent first question to pose should be: is there in these vignettes something unexpected or unforeseen to be handled by the actors involved, from which we can observe how they proceed to making sense of the situation?

Dismissing Entry#2

Taking a look at Vignette 1 we observe a 'business-as-usual' situation. The arguments put forward by Jurors A–D refer to two of the five evaluation criteria postulated in the competition brief, namely functionality and urban design (see above). Functionality is addressed by assessing an aboveground connecting wing as disturbing the sequence of rooms of the much-admired existing building (Juror D) – this argument is demonstrated by referring to *Plan#1* and *Plan#3* (Figure 5.5), which show that *Entry#2* proposes to open the outer wall of the Bonatz building on the first and on the second floor in order to connect the new building, thereby destroying the existing continuous sequence of rooms, which proved to be of high value for organizing (and experiencing) coherent exhibitions. Urban design is then addressed in two ways. First, Juror A, who refers to the use of wood as displayed on *Plan#1*, labels the connecting wing's façade as 'provincial' and therefore as unsuitable for the surrounding area. This argument is very likely based on the fact that wooden façades are uncommon in (the centres of) European cities but are rather common in suburban or rural single-family residential districts or industrial estates. Second, Jurors B and C (with Juror B laying ground for Juror C) stress the inadequacy of an aboveground connection – detached from the choice of material – for it 'interferes with the street as a defined space' (Juror C). This way of reasoning is completely in line with the information given in the brief (Figure 5.4). The jury's deliberation is predictable and does not show any surprising elements. Even Juror E, who is of a different opinion, raises his objections in such a tentative way – he tries to advocate the use of wood by referring to temporary architecture, which is highly popular these days, and he tries to trigger a discussion claiming that he sees a 'certain' quality (that is, a quality, which he cannot or does not want to determine) in connecting the two buildings aboveground – that the board's decision to dismiss *Entry#2* can be seen as unexciting general agreement. A quick decision without much ado – exactly what one would expect from a round of experts at this stage of the judgement process.

Reconsidering the proposal

The unforeseen happens in Vignette 2. Here, we observe an 'unexpected, or confusing event' (Maitlis and Christianson, 2014, p. 57) that interrupts the 'flow of things' (Weick, 1995, p. 100), that is, the flow of controlling the decisions made the day before. We use the term 'flow' since the board of jurors, having yesterday's minutes in front of them, smoothly reviews entry after entry confirming the decisions taken. All of a sudden, Juror F, while browsing the competition entry screening report, proposes reconsideration of *Entry#2*. What is not surprising though is the way he argues. The words 'Maybe we just need this entry in order to see more clearly what we are looking for. But maybe we should really reconsider this proposal' are frequently used among jurors in architectural competitions. If we additionally take into account that the jury members sit at a round table, out of sight of the entries, with each juror having an overview of all submitted entries in front of him/her, such a rather general statement, which draws a very rough comparison solely referring to *Entry#2*'s most obvious feature, is even less surprising. Juror F's statement adds little to Juror E's argumentation in Vignette 1. There is the notion of the 'wooden box' being a building in its own right (which is not new, but a direct quote from the text on *Plan#1*). His statement that *Entry#2* is 'the only entry that *convincingly* suggests an aboveground connection' again is of a general sort, and is a claim which he does not elaborate on. While in retrospect, this claim can be seen as an anticipated rejection of *Entry#16* (which is the only other remaining entry proposing an aboveground connection and is in fact rejected in the second round), at the time of its occurrence it has to be understood as Juror F noticing a 'cue' provided by *Entry#2* ('maybe we should *really reconsider* this proposition'). Yet, in the same way as Juror E in Vignette 1, he is unable to describe that cue. What is surprising is that the board of jurors accepts his motion. The decision to take *Entry#2* back into the competition is clearly not the outcome of an improved argumentation. We cannot really explain the board's reasoning in that respect. Maybe Juror F succeeds because of his strong reputation in the field, maybe some other jury members notice the cue as well – be that as it may, we are not concerned with this sort of speculation. What we argue is that this instance triggers a sensemaking process. The jury perceives 'the current state of the world […] to be different from the expected state of the world' (Weick et al., 2005, p. 409) – for they are about to reassess an entry that they unanimously rejected due to considerations that were established long before the competition started and have been an essential, clearly formulated part of the brief (Figure 5.4). Referring to Gioia and Gittipeddi (1991), one could say that Juror F's and Juror E's argumentation is sort of a 'sensegiving' in regard to the other jurors. It is exactly this sensegiving which Juror E alone did not achieve in Vignette 1.

Mobilizing cues and the irrevocable Entry#2

Looking at Vignettes 3 and 4, we can observe how the cues *Entry#2* – as a 'carrier of meaning' (Rafaeli and Vilnai-Yavetz, 2004a) – are deployed as 'actors negotiate new interpretations with other actors' (Stigliani and Ravasi, 2012, p. 1250). In particular, we will see how sense is constructed and spread and how cues that were neglected or overlooked and hence 'not available for sensemaking' (Weick, 1995, p. 52) at the very beginning are mobilized. The setting in Vignette 3 shows the jurors (after having evaluated all the other entries remaining in the second round) gathering in front of the plans and model representing *Entry#2*. After the jury acknowledges that *Entry#2* is the only one left with an aboveground connection, Juror F, pointing to Plan#3 (Figure 3.5) states that *Entry#2*'s connecting wing is not only a walkway from one building to the other but has a high functional quality: '[it] connects gallery spaces with gallery spaces via gallery spaces'. This thought constitutes a new perspective. Furthermore, referring to Juror E's argument in Vignette 1 where the wooden façade is linked to the idea of provisional architecture, Juror F calls *Entry#2* 'unpretentious'. Juror E immediately takes up this point claiming that a connection between the Bonatz building with its natural stone façade and the new building with its exposed concrete façade via an intermediate building with wooden cladding is an excellent approach to show that the connecting part is a serious space in its own right. Referring to Weick (1995) we can describe Juror F's and Juror E's actions as noticing and bracketing cues: an indeterminable intuitiveness is successively deployed until rational 'reasons' or explanations are found (Weick et al., 2005). Such rational explanations then have the power to spread. In our case, the slogan-like quality of these reasons may add to their performativity. Still, about half the jurors stick with their initial judgement as put forward by Jurors A and C. The exceptional situation of a jury (not counting the jury president) that is sharply divided into two equally sized opposing camps occurs and the only way to resolve the situation is for the jury president to ask for a vote by showing of hands. We note this, because in general juries are asked to debate as long as necessary, that is, as long as it takes to establish a broad agreement – a vote by showing of hands is not a real option. This exceptional situation shows brilliantly that the 'flow of things' (Weick, 1995, p. 100) is still not back to normal. The board of jurors still struggles to 'resume [its] interrupted activity' (Weick et al. 2005, p. 409) of producing stable judgements.

Two weeks later then, we see the two opposing camps dissolve (Vignette 4): almost all jurors now are convinced of the advantages of the aboveground connecting wing that *Entry#2* proposes. Moreover, with the prominent exception of Juror C, all jurors abandon the perspective that the aboveground connection as proposed by *Entry#2* is to be regarded as problematic – neither in regard to the existing urban structure of the surrounding quarter and to the street as a defined space, nor to the sequence of rooms of the Bonatz building. The reasons developed (and the way they are formulated together with *Entry#2* as an important carrier of meaning) succeed in giving sense to nearly all jury members. What is characteristic and important to note is that Juror C's arguments, which were shared by many jurors in the first two assessment rounds, are not disproved. While the wooden cladding *Entry#2* proposes for its connecting wing might be altered without difficulty, the fact that a two-story bridge interferes with the street as a defined space cannot be refuted. By acknowledging the functional advantages that *Entry#2*'s aboveground connecting wing brings with it, jurors exclude or bracket out its potential disadvantages.

If we take Vignettes 1–4 into consideration, we can state that in a business-as-usual setting, the story for *Entry#2* would have ended with Vignette 1. The 'unnoticed cues' in the jury's first round create a stable, robust assessment – while Juror E notices cues but is unable to bracket them in a way that makes sense to his fellow jurors. In Vignettes 2 and 3 then, we can see how Jurors F and E succeed in bracketing cues that are unnoticed by the rest of the jury board. In particular, Vignettes 2 and 3 show the active part the cues provided by *Entry#2* play as they inform, steer and further the jury's sensemaking. Finally, Vignette 4 indicates that sensemaking in jury boards

100 *Jan Silberberger and Ignaz Strebel*

of architectural competitions is not a process in which a thesis (a certain perspective on the problem) is contrasted with an antithesis to make a synthesis, or higher conclusion, reconciling the conflict between thesis and antithesis, but a process in which ambiguous cues, different ways of bracketing and antagonistic reasons are treated equally (at least for a short while) before they are prioritized (again) and a new, stable arrangement emerges.

Conclusion

Just as we were finishing this book manuscript we read in the media about the opening of the extension building of the Kunstmuseum Basel (see Figure 5.6).

Looking back at the whole process and comparing the renderings of *Entry#2* – which was rejected only at the final stages of the judgement process and whose main characteristic was the above ground connection wing – with photographs of the realized building, the difference is very striking. Having the realized building in front of our nose and while following the discussion and celebration of this particularly well received building in the media, the competition is perceived as a linear, rational and predestined procedure. The foregoing analysis and description of a single project – which was substantially different from the winning entry but nevertheless gained the attention of the jury board and was kept in the running until the end – demonstrates, however, that we need other understandings of how competitions work. We have

Figure 5.6 Kunstmuseum Basel, new building.
Source: Courtesy of Kunstmuseum Basel. Photo: Julian Salinas.

shown that this is to be found in the exchanges of arguments, the ways people orient and point at plans and documents and how they manage to find a sense of something that is not immediately obvious. It would be easy to denounce these informal, often not consciously noticed and perceived as unimportant details as problematic and 'corrupt' moments of the competition business. However, as our discussion shows, such conversations display something of the core aspects of what the qualities of competitions are. It is the unpredictable progress of the negotiations and discussions that determines the quality of a project and connected with this the assessment criteria that have to be fulfilled. Competitions and in particular jury sessions are not necessarily linear processes, but unfold with strategies of play, contingency and possibility. We would go as far as to say that a competition is successful, when it deliberately cultivates an understanding of dynamic progression and definition of the project to be built. It is obvious that from the perspective of the entire building process and not least of the construction industry such wayward and unpredictable moments are associated with risk and uncertainty. The analysis of the vignettes opens up a dilemma. On the one hand, it is obvious that such moments are far from arbitrary leading to hazardous decisions. We have shown that they trigger processes of enquiry, inspection and contextualization, so that resulting decisions are far from being simple subjective statements but must always be seen as co-constructed, judged and evaluated by a group of experts that are highly knowledgeable about the building process. On the other hand, the knowledge generated within the competition, can only be of use to the public in general and the building process specifically, if it comes through without these specific and sometimes erratic moments mentioned or even explained. And although it is helpful for research purposes to enter the jury room to understand the essential workings of conversations and how decisions are accomplished, for a competition to proceed successfully it is necessary that these instances and stories remain within the group, and the public as much as the client trust that the group of judges are able to select the best solution. It is a fundamental principle of contemporary competition practice that the jury board, after deliberation, explains which project it has selected and due to which reasons. How the decision has been reached must, however, remain confidential and should never be communicated. This sounds simple to do, nevertheless an agreement on this issue of all actors involved is something that every competition has to achieve if it does not want to be threatened by controversy.

References

Balogun, J., and Johnson, G. (2005). From intended strategies to unintended outcomes: the impact of change recipient sensemaking. *Organization Studies*, 26, pp. 1573–1601.

Chupin, J. P. (2011). Judgement by design: towards a model for studying and improving the competition process in architecture and urban design. *Scandinavian Journal of Management*, 27(1), pp. 173–184.

Clark, A., and Chalmers, D. (2010). The Extended Mind. In R. Menary (Ed.), *The Extended Mind*. Cambridge: MIT Press.

Gioia, D. A., and Chittipeddi, K. (1991). Sensemaking and sensegiving in strategic change initiation. *Strategic Management Journal*, 12, pp. 443–448.

Gioia, D. A., and Thomas, J. B. (1996). Institutional identity, image and issue interpretation: sensemaking during strategic change in academia. *Administrative Science Quarterly*, 41, pp. 370–403.

Harrison, A. (1978). *Making and Thinking: A Study of Intelligent Activities*. Indianapolis: Hackett.

Kreiner, K. (2010), Architectural Competitions – Empirical Observations and Strategic Implications for Architectural Firms. In M. Rönn, R. Kazemian and J. E. Anderson (Eds), *The Architectural Competition*, Stockholm: Axl Books.

Kreiner, K., Jacobsen, P. H., and Jensen, D. T. (2011). Dialogues and the problems of knowing: reinventing the architectural competition. *Scandinavian Journal of Management*, 27(1), pp. 160–166.

Maitlis, S., and Christianson, M. (2014). Sensemaking in organizations: taking stock and moving forward. *The Academy of Management Annals*, 8, pp. 57–125.

Orlikowski, W. J., and Scott, S. V. (2008). Sociomateriality: challenging the separation of technology, work and organization. *Academy of Management Annals*, 2, pp. 433–474.

Rafaeli, A., and Vilnai-Yavetz, I. (2004a). Instrumentality, aesthetics and symbolism of physical artifacts as triggers of emotions. *Theoretical Issues in Ergonomics Science*, 5, pp. 91–112.

Rafaeli, A., and Vilnai-Yavetz, I. (2004b). Emotions as a connection of physical artifacts and organizations. *Organization Science*, 15, pp. 671–686.

Schön, D. A. (1983). *The Reflective Practitioner: How Professionals Think in Action*. London: Temple Smith.

Söderström, O. (2000), *Des images pour agir. Le visuel en urbanisme*. Lausanne: Payot.

Stigliani, I., and Ravasi, D. (2012). Organizing thoughts and connecting brains: material practices and the transition from individual to group-level prospective sensemaking. *Academy of Management Journal*, 55, pp. 1232–1259.

Van Wezemael, J. E., Silberberger, J. M., and Paisiou, S. (2011). Assessing 'quality': the unfolding of the 'good'. *Scandinavian Journal of Management*, 27(1), pp. 167–172.

Van Wezemael, J.E., and Silberberger, J. (2015). We have never been 'Swiss' – Some Reflections on Helvetic Competition Culture. In J.-P. Chupin, C. Cucuzzella and B. Helal (Eds), *Architecture Competitions and the Production of Culture, Quality and Knowledge*, Montreal: Potential Architecture Books.

Weick, K. E. (1988). Enacted sensemaking in crisis situations. *Journal of Management Studies*, 25, pp. 305–317.

Weick, K. E. (1995). *Sensemaking in Organizations*. Thousand Oaks, CA: Sage.

Weick, K. E., Sutcliffe, K. M., and Obstfeld, D. (2005). Organizing and the process of sensemaking. *Organization Science*, 16, pp. 409–421.

Yanow, D., and Tsoukas, H. (2009). What is reflection-in-action? A phenomenological account. *Journal of Management Studies*, 46, pp. 1339–1364.

6 Jury board at work
Evaluation of architecture and process

Peter Holm Jacobsen and Andreas Kamstrup

Introduction

In 2006, the Carlsberg Group decided to move their brewery activities to a different part of Denmark. The old brewery area – the future Carlsberg City – is located in the city of Copenhagen and is therefore an attractive location for development. In the coming years, Carlsberg City will be turned into a multifunctional housing, business and recreational area in Copenhagen. This chapter builds on an ethnographic study of an architectural competition organized by a private client to develop and select both a design and a design team for a large building in the future city. The chapter focuses on the evaluation process during a new form of architectural competition. During three workshops held at Carlsberg, four teams presented and discussed their design proposals in front of a jury board. Each team presented their work separately from the other teams and after each presentation the jury board would ask questions to that particular team. The teams were also invited to ask questions to the jury board. The winning team was awarded the right to develop their design in collaboration with the client organization and the users.

Therefore, the question we seek to answer in this chapter is: *what happens when architects compete on both architecture and procedure?* In order to answer this question, we examine how the relationship between the assessment criteria from the competition brief and the jury board's professional judgements develops in the above-mentioned dialogue between the team and the jury.

Designing buildings in the upcoming Carlsberg City is considered a prestigious design task.[1] The particular competition format was chosen by the client organization, primarily because they worked under time pressure – a university college with 10,000 students was to use most of the building within a short timeframe if the project was realized. Therefore, the client organization wanted to select not only a team that would deliver the design promised in the competition, but also a team they could collaborate with within the timeframe and budget. In order to test and reward collaborative priorities and skills, this novel competition setup was introduced.

The competition was called a 'process competition' by both the client organization and the Danish Association of Architects, where the latter worked on developing, organizing and managing the competition together with the client organization. This specific competition procedure is not listed as an official procedure on the Danish Association of Architects homepage and the authors do not know of any other competitions in Denmark that follow this specific procedure. The participating teams and the members of the jury board had no prior experience of participating in such a competition setup because it was the first time that such a process competition had been organized in Denmark. However, the use of dialogue as an integral part of architectural competitions is becoming more common and the findings we present in this chapter are relevant to the further advancement of process-based competitions.

The chapter contributes to a better understanding of how the use of dialogue establishes social interactions and possibilities for participation between the teams and the jury board in the competition. It investigates the dialogue-based evaluation of the design proposals according to different criteria during the workshops. Our study supplements research on the work of competition juries in architectural competitions (Kreiner, 2012; Silberberger, 2012; Wezemael et al., 2011) when we show how the assessment criteria both develop and change in the negotiations at the workshops. The negotiations of the meaning of the assessment criteria are closely linked to the competition's solution space, because the assessment criteria evolve when evaluation and dialogues converge at the workshops. Evidently, this affects both the jury board's judgements and teams' work.

There is a particular focus on how the teams deal with a practical dilemma they encounter when presenting their work at the workshop: on the one hand, the teams can present and visualize their design at the workshop and get comments and feedback from the jury board, but on the other hand, the dialogue with the jury board cannot lead the teams to find the right solution to the problem since the criteria for selecting a winner is partly developed during the four parallel and at the same time isolated presentations given by each team. In these workshops, teams are confronted with difficult choices from the feedback they get on their presentation because of conflicting assessment criteria. We show how the teams' conditions for understanding (and learning about) the design task are difficult when the selection criteria are being negotiated during the process competition.

Structure of the chapter

In the following section, theories of evaluation and judgement in architectural competitions are introduced together with a situated perspective on visualizations and plans. First, we introduce this situated perspective to understand how a problematic situation takes place at a particular workshop. Second, we present our case study that gives insights into the setup of the process competition, the timeline and the assessment criteria described in the brief, and a short description of the development of Carlsberg City. Third, we present our methodological approach and the collected data. In the fourth section, we analyse one particular presentation and the subsequent jury board discussions and evaluations. The analysis focuses on the dilemmas and conflicts that evolve in the problematic situation. The chapter concludes with a discussion of the implications of the analysis and how the introduction of dialogue becomes both a source of creativity and a limitation of the solution space.

Evaluation in architectural competitions

From empirical studies (Kreiner, 2012; Kreiner et al., 2011), we know that competition criteria are not given a priori because preferences within the competition jury develop during the competition process. Empirical studies of the jury's work in four Swiss architectural competitions (Silberberger, 2012) also found that some evaluation criteria are not given a priori; rather, the criteria are found and developed when the jury discusses and evaluates the design proposals (Wezemael et al., 2011). We also know that the criteria used to select a winner are grounded in professional norms and intuition (Kreiner, 2012; Kaizeman and Rönn, 2009).

Economic sociologist David Stark (2011) argues that the architectural competition is an example of John Dewey's pragmatic understanding of value, because the principles of evaluation are found during the valuation process. Furthermore, Stark reminds us that judgements performed in organized competitions are different from judgements performed in head-to-head competitions and contests. The winner in a head-to-head competition, such as a football match or athletics meeting, can be measured according to a given set of rules: who scores the most goals or runs the fastest. Judgement in architectural competitions is different from head-to-head competitions and contests because designs in architectural competitions are evaluated according to several – often conflicting – judgement criteria.

Stark (2011) notices that evaluation in architectural competitions shares a number of aspects with competitions for national research grants, where a scientific review panel uses scores and ranks the research proposals according to criteria that emerge during the jury's deliberation. Kristian Kreiner (2012) has conducted empirical studies of jury work in a dialogue-based architectural competition. Kreiner mainly focuses on the jury's work and not the jury's interaction with the teams in dialogue-based competition. He observes how the criteria that are used to select the winning design are grounded in the jury members' professional judgement. Kreiner notes that the designs are difficult (or rather impossible) to rank, because the design task by nature is ill-structured (Kreiner, 2012, p. 411), and describes the relationship between evaluation and judgement in the following way:

> They are design proposals produced from a personal view, a unique interpretation of the design tasks developed over time in a sequence of judgments that form the attention and understanding of the salient dimensions of the task and their interrelationships. Design proposals produced in such a manner cannot be evaluated and compared analytically and objectively, since worth and attractiveness of a particular reading for the task must involve judgment.
>
> (Kreiner, 2012, p. 411)

Kreiner's (2012) argument supplements another aspect of the judgement process that springs from the understanding of what Cohen, March and Olsen (1972) call the garbage can decision making process, when he shows how the winning design is selected according to the professional intuition of the architects in the jury. The professional members of the jury (the architects) intuitively recognize the winner before the final meeting with the jury where the official decision is made. An important aspect in the competition jury's work is therefore to legitimize the final decision in relation to the assessment criteria when the intuitive judgement comes before the final decision. Kreiner reveals the complexity that is related to evaluation in the architectural competition when he shows that the competition criteria are being developed during the competition, while some aspects of the evaluation process are grounded in stable intuitive judgements made by members in the jury.

Earlier studies have investigated the interactions between jury boards and teams during competitions by focusing on the relationship between learning and decision making (Kreiner et al., 2011) or the role of sustainability assessment tools (Georg, 2015). Our focus on these interactions between jury board and teams are centred around understanding how dialogues influence the knowledge of, and subsequently the decisions by, the jury board. The dialogue in the isolated workshop creates dilemmas in handling asymmetrical knowledge when the jury board accumulates knowledge.

As mentioned above, the client organization want to see how the team's work and therefore the competition's evaluation criteria are different from the traditional open architectural competitions (Kaizeman and Rönn, 2009). The assessment criteria described in the competition brief are important in the process competition, but these criteria do not have the same substance as in anonymous architectural competitions.

In anonymous competitions, an important principle is that the competition jury does not know the identity of the architects behind the entries (Kaizeman and Rönn, 2009). In the process competition, the client organization *wants* to know the identity of the architects and, furthermore, how the teams present their work. In Jan Silberberger's empirical study of the jury's work in four Swiss architectural competitions (mentioned above), the jury uses assessment criteria that are written in the brief when they evaluate the entries that are submitted anonymously in invited project competitions (Silberberger, 2012, p. 261). In these competitions, the jury only interacts with the entries when they evaluate.

As mentioned previously, the teams are also evaluated on their ability to collaborate (i.e., their ability to participate in a dialogue, to be able to pose meaningful and relevant questions to the jury board, and to incorporate feedback in a short time frame between the workshops). We see a situation where the jury not only passes judgement according to the assessment criteria that are written in the competition brief (Silberberger, 2012) based on the jury members' capacity to draw on intuitive judgement (Kreiner, 2012) and their general knowledge about the entire construction project, but also according to the verbal and processual skills of the particular architect team. This performative judgement is not only passed based on objective criteria or intuitive and general knowledge but, furthermore, it also becomes relational between the performing actors: the jury board has to decide which team delivered the best presentation and shows the best ability to participate in the dialogue.

In the following, the competition process will be considered as a social practice, where the evaluations of the presentation are situated around the discussion of the architect team's visualization of the building and the subsequent discussions. This theoretical approach builds on the concept of situated learning originally developed by Jean Lave (1988; see also Lave and Wenger, 1991), whose focus is on how people learn in practice. Recently, the situated perspective has contributed to an understanding of how professionals – such as architects – make judgements in practice (Styhre, 2013).

In this study, the focus is on how the teams and the jury board in the competition are gathered around discussions about a common cause (Axel, 2009) to develop and select the design of the building by judging and selecting a design for the building. The discussions and negotiations in the workshop about the design of the new building are understood as conflictual because the participants have different subjective perspectives on the common cause (Axel, 2009). The jury board's on-going evaluations are grounded in different understandings of the visualizations that are related to, for example, economy, functionality and aesthetics.

At a given workshop, the particular team visualizes how the building can be designed and how the team will collaborate with the client organization in the next phases of the construction project. Therefore, the material aspects play a central role. The PowerPoint presentations and pictures the team use as visual representations are understood as an integrated and active part of the social practices. In practice, the visualizations are a part of the plans and strategies of the architect teams and we understand these visualizations in line with Lucy Suchman's (1987) situated perspective on plans. She argues that people take the world for granted in their everyday lives, but when the obvious and taken-for-granted becomes problematic, people have to represent and use plans in the practices they are a part of. Representation can occur before a 'problematic situation' (i.e., when the teams make a plan before presenting it to the jury board). But representation can also take place after a breakdown because a situation at the meeting did not make sense (Suchman, 1987, p. 52). Therefore, plans do not determine people's actions – plans are resources for actions in problematic situations (Suchman, 1987). Below we present our case in a more detailed manner, allowing the reader to follow our arguments in the analysis and eventually come to the same conclusion.

Case presentation

The Carlsberg City competition setup

The process competition was centred around the organized dialogues between the four invited teams and the jury board. The jury board was comprised of fifteen persons, with six acting as jury members and nine as advisors. Two of the jury members were appointed by the Danish Association of Architects. The jury board also consisted of representatives from the client organization, the municipality of Copenhagen, the future users, and the competition managers from the Danish Association of Architects. The competition process and progression is illustrated below (Figure 6.1). As mentioned above, the workshops are of key interest: here the teams had the opportunity to ask the jury board questions during their presentation, and at the same time the jury board asked the teams questions. The teams had one week between each workshop presentation to rework proposals and presentations.

According to the assessment criteria described in the brief, the four teams were evaluated according to 1) how they visualized the design of the future building, 2) how they would collaborate with the client organization if they were selected, 3) whether the design could be realized within the given time frame and budget and in a way where all functional demands were met, and 4) fee. The explicit focus on collaboration and teamwork was related to both the ability of the team to work together with the rest of the client organization, the future users and stakeholders.

The members of the jury board participated in all workshops and meetings and therefore they had knowledge about how all four teams worked on their designs. The four teams only participated in their own workshops and the panellists in the jury board could not give ideas and knowledge from one team to another. Some of the teams were nervous about their ideas being given to the other teams during the process and they explicitly asked if they could be sure that their ideas and questions were not given to the other teams. In this particular competition setup, the jury board discussed and negotiated different aspects of the brief with the four teams.

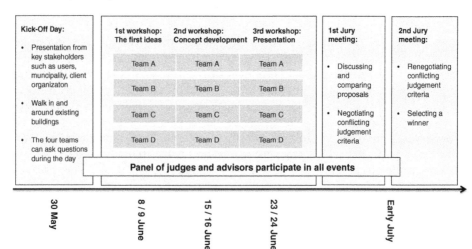

Figure 6.1 Timeline.
Source: Authors.

In the following, we will introduce the most relevant ideas in the master plan for Carlsberg City. It is important to keep these ideas in mind when seeking to understand why the competing architect teams act as they do.

Developing Carlsberg City

In 2006, Carlsberg organized an open international architectural competition to find a master plan for the new city area. Two hundred and twenty-one proposals were handed in and the winner was the Danish architect firm Entasis with a proposal called 'Our space'. On the one hand, the master plan is inspired by the small-scale classical city houses found around Copenhagen, but it also introduces a series of towers, which is a new aspect in Copenhagen where the city centre's skyline has been strictly regulated with limitations on building height.

One of these towers (see Figure 6.2) is located and grounded in the building that our teams are developing designs for. In line with the client's vision, the master plan aims to create a city that is multifunctional by mixing different forms of dwelling, educational institution, shop and cultural institution. In 2009, the master plan won the prize as the best master plan at the World Architecture Festival in Barcelona. In the years following the open competition, the master plan has been reworked into a number of district plans for the area.

The process competition was organized to produce concrete visualizations of one of the first buildings in Carlsberg City. The building, at more than 80,000 square metres, is located next to a central railway line in Copenhagen and has a budget of 1.3 billion Danish kroner.

The left side of the building – 'The Hanging Gardens' – is listed for preservation and therefore the architect teams needed to integrate them into their designs. During

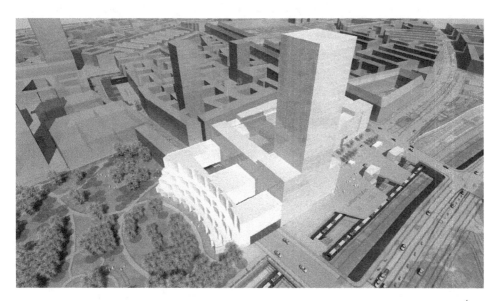

Figure 6.2 Visualization of the building from the competition brief.
Source: Carlsberg City District.

the process competition, the client organization negotiated with future users of the building, including a new university college that wanted to establish a campus for 10,000 students. Therefore, the process competition was a way to visualize how the university college could be a part of the future building. The university college had representatives participating in the workshops because they represented such a large potential user group.

Methodology and data collection

The analysis is based on a single case study (Flyvbjerg, 2006) and focuses on the evaluation of team A's presentation at the first workshop. Our collected data consists of observations of all meetings, workshops and jury meetings (see Figure 6.1), semi-structured interviews with key informants, in situ interviews, more than 200 digital photographs from workshops, audio recordings from all twelve workshops and the four teams' PowerPoint presentations for each workshop. The data collection started approximately one month before the competition began, with observations of meetings in which the client organization worked on the brief and planned the workshops. Before each meeting, the researchers were provided with agendas and access to work-in-progress documents of the brief in a shared online folder (Dropbox).

The aim of using different methods to investigate the competition process was to understand how dilemmas and conflicts were situated in social practices when the teams were presenting their work at the workshops. The methodological approach to the field is also inspired by Geertz's concept of 'thick descriptions' that has been used to understand aesthetic knowledge as a part of design work (Ewenstein and

Whyte, 2007). Before and during the competition, five semi-structured interviews were conducted with key informants. After the jury process ended, seven semi-structured interviews were conducted with representatives from the client organization, user representatives, and leaders from two architect teams and two judges. The semi-structured interviews were structured around themes and dilemmas that were observed during the workshops.

Our study is based on the data collected from the first workshops – the 'first ideas'. The following investigation is based on an analysis of the interactions between the jury board and team A at the first workshop (see Figure 6.1). Data recordings, digital pictures and the team's PowerPoint presentations were used to analyse the dialogues between the jury and the team.

Analysis: inside the competition

We examine a particular instance: the very first meeting between architect team A and the jury board. We divide the analysis in two: first, retelling how the team presented their plans and visualizations and, second, reproducing key elements of the discussion within the jury board. Combining these two instances will show how judgements unfold in the process competition and how the particular dialogues keep changing the solution space.

The team's presentation and visualization

Using a first instance of data recorded from the workshop, we will present two different aspects of the team's presentation that challenge the assessment criteria of the brief.

The first aspect of the team's strategy is that they suggest moving the tall tower to a corner of the building. The location of the tall tower is defined in both the master plan and the district plan. It is visualized in the competition brief (see Figure 6.2) and the teams are not expected to challenge this location but rather visualize the façade. Therefore, the proposal goes against the competition brief. The team chose to use visual representations of the building in their presentation and they use the workshop to visualize the moving of the tower based on pictures in their PowerPoint presentation. The team argues that the tower is placed on a base that is not big enough and that the tower will not be slim enough with the placement in the master plan and district plan. The team explains to the jury board that moving the tower will create some problems in relation to turbulence and light, but also that they are working on solving these issues. The team argues that moving the tower will give value to the new functions in the building and valorize the square in front of the building. If the tower is moved to the corner of the building, it will also reflect light on the new square at Carlsberg Station (see Figure 6.2). Furthermore, the team shows pictures of different models of the building from the material in the brief and illustrates three different positions of the tower for the jury board. In short, the team is using the workshop to suggest and visualize that the tower should be moved and, furthermore, that they would like to discuss this suggestion with the jury board. This example illustrates how the process competition allows for raising questions and suggestions that go against the brief.

The second aspect is about preservation. The team wants to preserve as much of the old brewery building as possible in the design of the new building. Their argument for preserving the building is that they do not think it is possible for architects to design the architectural qualities that are a part of the old brewery building. The team shows pictures of other old buildings that have been transformed where the old buildings have been preserved in new buildings as a part of their argumentation and presentation. The team also suggests reusing materials such as stone, copper and steel from the existing building.

To sum up the team strategy: a number of aspects in the team's presentation challenge the premises in the competition brief, the master plan, and in the district plan by suggesting that some central functions be moved and that the existing building be preserved. The communication in this first part of the workshop is primarily one way: the team presents their visualizations and the jury board watches and listens. The presentation gives the jury board the first impressions of team A's architectural design (first assessment criteria) *and* the team as possible collaborators (second assessment criteria).

However, this strategy creates new problems and dilemmas that are related to the design of the building when the jury board discusses and evaluates the presentation. We have a problematic situation because the taken-for-granted is challenged in the team's representation of the building (Suchman, 1987).

Jury board at work

Directly after their presentation, the team leaves the room and the jury board is left with a number of questions that they have to find answers to because they have to give the team feedback ten minutes later. The two mentioned aspects are difficult to answer because the questions touch upon very central issues in relation to the competition brief. The discussion reflects that the members of the jury board have worked with the development of the master plan and have spent several years planning the new city. Some aspects of the evaluation are similar to the evaluation process described in anonymous competitions (Silberberger, 2012), but as already mentioned, aspects became relevant during the team's presentation that challenged the criteria of the competition brief. In anonymous competitions, entries that violate the specifications of the brief cannot normally win a competition (Silberberger 2012), but in this process competition, the team can test their ideas before they present their final proposal.

Knowledge about the project history is important when the jury board evaluates the presentation. The members of the jury board take explicit formulations in the brief into account in their evaluation. However, all the arguments about why the tower is placed where it is in the building are not described in the competition brief. The members of the jury board draw on their knowledge about the project when they make their judgements. This aspect of the evaluation is understood as being in line with what Kreiner (2012) describes as a sequence of judgements based on the jury board's professional knowledge and intuition. The panel members have to draw on such intuitive judgements when they deliver answers to the team within the short time frame.

Another important issue relating to the jury board's judgement is the preservation of the existing building. Several members of the jury board agree that the existing building is unique, but the members of the client organization do not think that it is possible to achieve the 80,000 square metres in the new building if the old brewery building is preserved. Therefore, the question of preservation opens up concerns for the client organizations regarding economic and functional feasibility (third competition criteria). Below, we reproduce some of the actual conversation between the jury board panellists. We focus on examining how the team's presentation of the problematic situation was dealt with.

Client advisor: I agree that constructing two basements below an already existing building contains a lot of challenges ... and so you would have to examine what could be done instead. If you [the team] convincingly can argue that you [the team] would be able to solve the issues – the wishes – then everything is fine ...
Project manager: Yes, yes.
Client advisor: It is an exercise we've been through, realizing that we did not ourselves have the capacity to get the amount of square metres necessary in the existing building, and then the project simply falls apart ...
Jury member 1: That is a splendid attitude. If it can be formulated like that to them [the team], then ...
Project manager: I agree ...
Client advisor: But are they [the team] able ... everyone thinks this building is very beautiful, it is not like we do not want it. What we have not been able to realize ... as soon as money is involved the project falls apart.

The client organization and their advisors have been working on a way to preserve the old brewery building, but it has not been possible to find a solution where all demands are met. Based on their evaluation of the presentation, the members of the jury board agree on telling the team that they have been working intensively on preserving the old building, but that the client organization has not been able to find a solution. Based on the team's presentation, the members of the jury board discuss the solution space based on the premises in the brief. The members representing the client organization are open enough to reconsider this solution space.

However, a new problem arises that is related to the question of preserving the old brewery building: can the team preserve the old building so that it does not collide with the assessment criteria in the brief? The problematic situation affects the jury board's work. The jury board has to renegotiate the meaning of the assessment criteria before they meet with the team again. To do this, the jury board discusses how much of the building the team wants to preserve, again based on their presentation. It is not clear from the presentation whether the team wants to preserve the entire building or just parts of it. The jury board has to take a look into the actual formulations in the competition brief.

Competition advisor (Reads for the rest of the jury board): Ok! Expected to be torn down!
Several advisors: What do you say?
Competition advisor (Reads again): The brewery building including the train building can be expected to be torn down!
Several advisors: That's an opening!
Jury member 2: Then they have a challenge with the square metre!

In the brief, it is written that the old brewery building is expected to be torn down. In the district plan, it is written that some of the old building's façade can be preserved. If the team chooses to proceed with their strategy of the building being preserved then they have a challenge concerning the number of square metres available, as jury member 2 states.

The team enters the room again and the jury board tells them that the building has to be at least 80,000 square metres for the project to be realized within the economic framework. The team announces that they are looking both above and below the existing building to find these square metres. One of the problems with preserving the old building is the large amount of water under the construction site. One of the client advisors tells the team that it is a problem to dig into the water and that the geologists would have to 'rewrite' the history books because the soil under the construction site is as hard as concrete. Such hard soil has never been seen before.

Concluding discussion

We have described and analysed the evaluation process in a new form of architectural competition. Our case analysis has shown how evaluation in the process competition takes place in the interaction between the team and the jury board in an early phase of the process competition. We argue that this evaluation is crucial in order to understand the architectural solutions that are developed in the competition. Furthermore, we argue that assessment criteria are not given a priori, rather the competition criteria are developed and negotiated in the competition process. Our analysis showed how team A's visualizations created a problematic situation which the jury board had to take into account in their evaluation and feedback to the team. A new understanding of the assessment criteria emerged from the jury board's negotiations just after the team's presentation. In the evaluation, the jury board assessed the presentation according to the criteria formulated in the competition brief, but also according to their broader knowledge about the Carlsberg project. The team received feedback based on this new understanding and interpretation of the assessment criteria by the jury board.

The analysis of the team's presentation has shown that a number of problems and dilemmas have to be faced when the team challenges aspects of the competition brief. In one way, it is perceived as positive when the team challenges the brief and raises 'tricky questions', as one jury member formulates it in the feedback to the team, but this challenging strategy also created dilemmas that the team had to consider before the next workshop.

In the process competition, the team can test ideas that would violate the programme in anonymous competitions (Silberberger, 2012). In the dialogue between the teams and the jury board the solution space can be negotiated before the teams present their final proposal. But the dialogues with the jury board at the workshops raise new questions. After the workshops in the process competition, the teams reconsider their plans after getting feedback from the jury board. In the situated perspective, it is not the essence of the plan that is interesting (Suchman, 1987) and this is also the case in the process competition. Rather, it is how visualizations are used in practice and what relationships they have with other social and material plans, such as the master plan and district plans for the entire Carlsberg City, that become relevant.

The suggestion of moving the tower illustrates how the team's visualizations of their plan creates a dilemma because the location of the tower is connected to the jury board's knowledge about the development of Carlsberg City. The team tells the jury board that they are already working on the consequences of moving the tower: at the workshop, they explain that it creates problems with turbulence and reflection of light on the square around the building if the tower is moved, but the team is surprised when the jury board tells them that they have been working on moving the tower for months and have been struggling with a strict regulation of the placement of the tower in the city plan. Therefore, the question about moving the tower contributes to surprising feedback and knowledge for the team's work.

Our situated understanding of evaluation and plans (Suchman, 1987) opens an understanding of how conflicting views become present in practice when the team visualizes their design of the building, and that dialogues entail both possibilities and limitations in the competition process. When the team challenges how the building could be designed, they open up questions that relate to why the competition brief is formulated the way it is, and thereby challenge the basic criteria of the competition as formulated in the brief. When the team challenges the criteria and asks if they can preserve the existing building, the jury board have to discuss and negotiate the formulations in the competition brief. The dialogues within the jury board and the dialogues between the architect team and the jury board illustrate contradictions that are grounded in different concerns related to the building. In particular, the question of preserving the old building clashes with the requirements for the number of square metres in the building. When the architects suggest preserving the existing building, they are met with concerns that relate to functions and economics (competition criteria three). These concerns are again related to business plans for the building.

The team has to consider these contradictions and dilemmas in relation to their design work before they present at the next workshop. How is the team supposed to handle the dilemmas concerning moving the tower? Should the team stick to its plan or move the tower back again for the next workshop? Should the team proceed with developing a design of the building that integrates the old brewery building? It is not possible to find one correct solution to these questions, which is why the team's work is riddled with dilemmas.

Also, these choices that guide the team's presentations in the process competition are not just about the design of the building. They have to adopt the feedback and present again a week later at the next workshop and demonstrate that they have listened to the the jury board's feedback from the dialogues at the workshop. The jury board are already evaluating the team as potential collaborators. The process competition is also about demonstrating that the team is able to listen to what the future employer is saying. But how does the team adapt to feedback that is grounded in contradictions? Is the right strategy to stick to the plan and hope to convince the jury board or is it to follow the guidelines from the jury board and change their plan accordingly? More research is needed to understand the contextual dilemmas and problems that are part of new forms of architectural competitions such as the process competition.

We find it noteworthy to mention one last challenge in relation to interactions between the jury board and the teams: the size of the jury board. This aspect was discussed at a seminar in January 2016 concerning novel architectural competitions. We presented the case study and some of the findings that we have analysed in this chapter. Several participants at the seminar also participated in the process competition and they expressed that it was difficult to understand which comments and feedback counted in the final evaluation when they were confronted with a jury board that consisted of many members with different agendas. A current form of process-based competition that is often used in Denmark is a two-phase project competition. In these competitions, the jury only consists of four or five persons. One of the arguments for limiting the size of the jury board is to eliminate conflicting judgement criteria, since the team negotiates with only a few people.

Acknowledgements

We would like to thank all the participants in the process competition for letting us follow their work. We would also like to thank Kristian Kreiner, Ursula Plesner, Jan Silberberger, Ignaz Strebel and the anonymous reviewer for thorough readings and valuable comments.

Note

1 The competition was publicly announced on the Carlsberg City homepage and on the Danish Association of Architects homepage. Seventeen teams with five architect firms in each team applied to participate in the competition process. The four teams were selected by the client organization based on a short description of their team and CVs.

References

Axel, E. (2009). What makes us talk about wing nuts? Critical psychology and subjects at work. *Theory & Psychology*, 19(2), pp. 275–295.
Cohen, M. D., March, J. G., and Olsen, J. P. (1972). A garbage can model of organizational choice. *Administrative Science Quarterly*, pp. 1–25.
Ewenstein, B., and Whyte, J. (2007). Beyond words: aesthetic knowledge and knowing in organizations. *Organization Studies*, 28(5), pp. 689–708.
Flyvbjerg, B. (2006). Five misunderstandings about case-study research. *Qualitative Inquiry*, 12(2), pp. 219–245.
Georg, S. (2015). Building sustainable cities: tools for developing new building practices? *Global Networks*, 15(3), pp. 325–342.
Kazemian, R., and Rönn, M. (2009). Finnish architectural competitions: structure, criteria and judgement process. *Building Research & Information*, 37(2), pp. 176–186.
Kreiner, K. (2012). Organizational decision mechanisms in an architectural competition. The garbage can model of organizational choice: looking forward at forty, *Research in the Sociology of Organizations*, 36, pp. 399–429.
Kreiner, K., Jacobsen, P. H., and Jensen, D. T. (2011). Dialogues and the problems of knowing: reinventing the architectural competition. *Scandinavian Journal of Management*, 27(1), pp. 160–166.
Lave, J. (1988). *Cognition in Practice: Mind, Mathematics and Culture in Everyday Life*. Cambridge: Cambridge University Press.
Lave, J., and Wenger, E. (1991). *Situated Learning: Legitimate Peripheral Participation*. Cambridge: Cambridge University Press.
Silberberger, J. (2012). Jury sessions as non-trivial machines: a procedural analysis. *Journal of Design Research*, 10(4), pp. 258–268.
Stark, D. (2011). What's Valuable? In P. Aspers and J. Beckert (Eds), *The Worth of Goods: Valuation and Pricing in the Economy*. Oxford: Oxford University Press.

Styhre, A. (2013). *Professionals Making Judgments: The Professional Skill of Valuing and Assessing*. Basingstoke: Palgrave Macmillan.

Suchman, L. A. (1987). *Plans and Situated Actions: The Problem of Human–Machine Communication*. Cambridge: Cambridge University Press.

Van Wezemael, J. E., Silberberger, J. M., and Paisiou, S. (2011). Assessing 'quality': The unfolding of the 'good' – collective decision making in juries of urban design competitions. *Scandinavian Journal of Management*, 27(1), pp. 167–172.

7 Jury boards as 'risk managers'
Analysing jury deliberations within architectural competitions against the background of risk management

Camille Crossman

Introduction

In 1983, the Opera Bastille Public Corporation launched an international – anonymous – design competition for the new construction of the Paris Opera House at La Bastille. As underlined by Deyan Sudjic, architectural critic and director of the Design Museum of London since 2006, towards the end of their assessment of the 756 entries, the competition jury 'were under the impression that they had chosen the best of a lack luster field of submissions' (Sudjic, 2006). The project of their choice convinced them with its 'purified geometry' and 'purist white skin' (Sudjic, 2006), which reminded them of Richard Meier's High Museum in Atlanta (Sudjic, 2006). In fact, the jury saw such a strong analogy – also as regards the drawing technique – that they were almost sure that the project they were dealing with (and later selecting as the winner) was the work of Meier as well (Sudjic, 2006). As François Chaslin elaborates, it was not primarily the project that persuaded the jurors but their own assumption regarding its author. Meier, who was at his prime at that time, they further assumed, would significantly improve his proposition once he was awarded the commission (Chaslin, 1985). Yet, as can be the case with assumptions, this one turned out to be false. When the envelopes were opened, the jury discovered that the project they picked was not the product of the American architect with the proven track record and international renown they had in mind but was in fact penned by an unknown Uruguayan practicing in Canada named Carlos Ott.

Deyan Sudjic concludes his account stating that the Opera House 'remains a flawed project, even after the technical problems with its cladding system, and its malfunctioning mechanical stages had been resolved' (Sudjic, 2006). So did the jury pick the wrong project? Or was the competition entry a good proposition, which has been weakly executed due to an architect that lacked experience? Would Richard Meier have built a better Opéra de la Bastille? All of this is pure speculation. Which nevertheless brings us to the central topic of the chapter at hand.

In an architectural competition's jury, jurors do not only assess the projects as displayed on the plans, they also (try to) make meaningful predictions about its future. Assessing the entries of an architectural competition represents – in part – an exercise in speculation, which aims to anticipate or to predict their future performance. Thus, architectural judgement is also built on hypothesis and speculations. What will the needs of users of the project at sake be like in thirty-five years? Is the project addressing them? How is the project going to behave once implanted in its context (architectural, cultural, political, environmental, etc.)? Is the project flexible enough to stand the test of time? Will the space built according to the proposed scheme support the objectives of the project as defined in the brief? Will the architects, authors of the project, be flexible and easy to work with during the construction of the project? Will the unresolved aspect of the project be resolved through the execution phase? Will they be open and able to improve and modify to the original concept when necessary?

As observed during the fieldwork observations conducted for our thesis (for methodological aspects and selection of cases see section 3) such questions and preoccupations are regularly raised by an architectural competition's jury. Thus, statements like the following can be heard:[1]

> This project has its weaknesses. Just to take the example of the façade. Yet, all these weaknesses can easily be reworked.
>
> This entry proposes a very coherent, very strong concept. It helped us to sharpen our assessment criteria. But, let's face it: it is not feasible. We have to exclude it.

Such statements indicate again that, through speculative processes, juries also judge the potential – the capability, the feasibility – of an entry, and not just its actual properties. In notorious cases, such as the Opéra de la Bastille, juries speculate about the authors and even about the capacity of the architects as a future business partner, although the procedure is completely anonymous. Furthermore, these speculative comments or arguments directly participated in building a collective architectural judgement since the jury decided to rely on their assumption – as speculative as it was – to designate the winning project. In other words, in a competition's jury, speculative judgements are considered valid arguments leading to the selection or the elimination of a project. But how do jurors formulate, justify and validate their speculation? How do certain speculations become acceptable enough to validate a judgement and to make a decision? This is the problematic we explore through this chapter.

Concepts of risk and architectural judgement

Originally, speculation meant to observe, to reflect on something. In its general definition, speculation is now defined as 'Ideas or guesses about something that is not known' (Merriam-Webster online dictionary). Nowadays, speculation is an important concept in the stock market field, as it designates an 'Activity in which someone buys and sells things (such as stocks or pieces of property) in the hope of making a large profit but with the risk of a large loss' (Merriam-Webster online dictionary). This last definition is important since it introduces the idea of risk: when risks are unknown, solutions indefinite or future uncertain, speculation can occur to different degrees.

In his work *Risk* (1995), a fundamental text in the field of finance and risk management theories, John Adams defines and discusses risk's and speculation's fundamental principles in the modern and contemporary era. Although these concepts are mainly related to the stock market field, Adams also discusses their importance in the decision making process in any situation: politics, economic policies, health system, road infrastructures, security matters, and so on. Adams defines risk as follow:

> 'Risk' is defined, by most of those who seek to measure it, as the product of the probability and the utility of some future event. The future is uncertain and inescapably subjective; it does not exist except in the minds of people attempting to anticipate it. Our anticipations are formed by projecting past experience into the future. Our behaviour is guided by our anticipations. If we anticipate harm, we take avoiding action.
>
> (Adams, 1995)

This broader definition of 'risk' allows the linking of risk management processes to many aspects of the architectural field, in general, and to the architectural conception and judgement processes, in particular. An architectural project represents, through the conception phase, a series of complex exercises of projection of various intentions into a hypothetical future – as first imagined and described by the client's needs and desires. As 'project' is etymologically linked to the action of projecting – to project, which originally means to throw forward (Boutinet, 1990) – the idea of a future becomes intrinsically intertwined with the process of making a project, as an attempt to design, to predict – and eventually to literally build – that future. But since 'the future is uncertain and inescapably subjective' (Adams, 1995), architectural projects also represent risks, whether they be conceptual, economical, political, environmental, contextual, constructive, and so on. Indeed, the realization or the construction of an architectural project always involves unforeseen surprises, since the context – in its broad definition – is never completely predictable and since an infinite number of variables are entangled. Thus, these risks have to be identified and discussed, or managed, as precisely as possible, throughout the conception and the realization of an architectural project.

However, Jens O. Zinn, an Australian sociologist also specialized in the problematic of risk and uncertainties, underlines some important facts about the concept of risks in his book *Social Theories of Risk and Uncertainty* (2008).

> Nothing is a risk in itself; there is no risk in reality. But on the other hand, anything can be a risk; it all depends on how one analyses the danger, considers the event.
>
> (Zinn, 2008)

> Since risk implies an uncertain future, real dangers as well as concerns, fears, or imagination are indissolubly parts of risk. Risks are always real and constructed.
>
> (Zinn, 2008)

In fact, the task of the juror of an anonymous architectural competition is to judge the quality and the future performance of a project still at its conceptual and un-built state, without any other information than the boards and the texts submitted by the participants. Therefore, the jurors' challenge is to try to make sense (Weick et al., 2005) of these diverse and complex elements. To do so, they imagine the behaviour of the propositions through a series of hypothetical – speculated – projections on the project in relation to its context. For example, we regularly observed discussions regarding maintenance and weather issues: is there enough space for the snow removal truck? Are the courtyards sheltered from the wind? And so on. These speculations are built

120 *Camille Crossman*

with the jury's own judgement and their personal experiences to analyse and judge the projects, individually and collectively. As Zinn argues:

> risk is not only a technique for managing an uncertain future. It is part of societal discourses, which produce the knowledge to define reasonable action and decision-making. Thus risk is unsolvable linked to normative and moral issues [...]. Even when risk is introduced as 'evidence based,' as for example in criminology or medicine, whether and how such 'objectivist' calculations are applied and which criteria are selected are necessarily confounded with moral judgments.
>
> (Zinn, 2008)

Thus, considering the example cited in our introduction, how can we relate risks and risk management theories to architectural judgement theories? How can risk management theories provide a better understanding of competition juries' decision making processes and contribute to a better understanding of architectural judgement? Here, we pose the hypothesis that architectural competition juries represent the perfect scientific situation to observe and enquire into these questions and that risk management theories represent a pertinent theoretical framework to provide some hypothesis.

Case selection and methodology

Through the different phases that constitute the architectural design process, the judgement process as performed in an architectural competition's jury appears to be an ideal window to enquire about architectural judgement through the lens of risk management.

Three types of procedure with regard to a project competition are generally used: the open, the restricted and the invitation procedure. An 'open' procedure means that every architecture office is allowed to submit, anonymously or not, depending on the decision of the organizer of the competition, solutions to the given task. In contrast, within an invitation procedure the client / organizer asks a rather small number of practices to anonymously (or not) compete against each other. Open competition and competitions following an invitation procedure can be held in one or two stages of design. In a one-stage competition, a winner is selected among all the submitted propositions. In a two-stage competition, a certain number of finalists (usually three to seven) are selected to further develop their concept through a second stage of design. It is only at the end of the second stage that a winner is selected. The restricted procedure is a sort of mixture of these two types. In a restricted competition every architecture office is allowed to apply on a first level by submitting (usually two) reference projects taken from its portfolio (these reference projects are preferably closely related to the competition's task) and by completing an application form. Then, in the so-called pre-qualification jury assessment session, the board of jurors evaluates these reference projects as well as the practice's general track record. On this basis it invites a certain number of offices to develop (anonymously submitted, or not) designs for the problem at hand. Hence, in a restricted as well as in an invitation procedure the client knows which architectural offices handed in projects but may not know the projects' authors when the designs are submitted. It is important to underline that in Canada, unlike in the European countries, competition procedures are not mandatory even for public building projects. Moreover, when a competition is held, clients and professional advisors are allowed to decide which procedure they will adopt and whether their competition will follow an anonymous procedure or not.[2]

Over several years we observed four Canadian competitions' juries following various types of procedure: open or restricted, anonymous or non-anonymous, held in one or in two phases of design, and so on. Since the juries are always conducted behind closed doors, access was granted under strict conditions: the data collected had to remain anonymous. To do so, we carefully noted every intervention of each juror by hand (no recording material was permitted) during each observation in order to produce as verbatim field notes as possible of the judgement processes. Afterwards, these notes were cleared of any names or details that could compromise the anonymity of the members. We submitted observation requests for each architectural competition organized in Canada, and during the period of our research we obtained four permissions. Therefore, our data is not based on representativeness of the cases but rather on opportunity of data collection. Following constructivist methodology approaches, we then compared the discussions of the judgement process of observed juries. We were then able to identify patterns, similarities, contradictions, and so on and to highlight several problematics regarding architectural competition juries' situations and architectural judgement in general.

In order to investigate the specific problematic tackled in this chapter, we selected a specific instance from one of the observed competition's jury. To discuss the findings of our case studies – and to respect the strict rules of confidentiality in effect in a jury – these extracts are presented in two vignettes. The term 'vignette' in ethnography refers to a short concentrated self-contained and anonymized description of a situation. A vignette is not to be mistaken for a transcription (e.g., of a tape recording). It is in fact a reconstruction of a situation based on field notes. The way of working with vignettes is borrowed from Söderström (2000) and Silberberger (2012; 2011), who consider their vignettes to be well-understood as fabrications and selections of their observations that do not intend to 'objectively' reproduce the observed situations. These vignettes represent the source of our interrogations as well as constituting the basis of our analysis and findings. Finally, a few details of the competition are intentionally left unknown for two reasons: first, in order to respect the anonymity of the procedure, and second, in order to highlight the discourses, the argumentation strategies and the fact that these discussions could be observed in any competition jury's process.

Behind closed doors: competition juries' processes

The discussions presented in the two vignettes took place during the first stage of a two-stage multidisciplinary competition (architectural, urban design and landscape). The programme of the competition concerned the revitalization of an Olympic Park built a few decades earlier.[3] The programme concerned urban revitalization, densification of the area, new housing, new public services such as a new public library, development and protection of the existing green areas, and so on. This design and build competition was open at the international level and followed an anonymous procedure for the two stages of the competition. In the first round, fifty-two projects were assessed. The goal was to select five to seven projects as finalists for the second stage of the competition. The first stage's judgement procedure was spread over two full days of debates. The jury was composed of eleven jurors allowed to vote, of

122 Camille Crossman

which six were architects. Each multidisciplinary team had to submit a concept or a vision for the project. Two A1 boards containing plans, section and 3D rendering were required. For half of the judgement procedure, the technical experts presented the projects to the members of the jury. They had previously analysed in detail many technical aspects of the projects: built and green surfaces, estimated cost, traffic studies, and so on. For the second half of the procedure, the jurors were standing, discussing and voting in front of the boards.

One particular aspect of that competition's judgement process was the presence of technical experts – not members of the jury and not allowed to vote – during all the discussions. While a few silent observers are generally present during the jury's meetings to vouch for the procedure's transparency, we rarely observed the presence of technical experts during the judgement process. Furthermore, in this particular case, the technical experts were also allowed to discuss with the jurors and give their opinion throughout the whole procedure.[4]

Vignette 1

A: This project is full of mistakes. The courtyards are too small, the buildings are too close to each other, the graphics are too garish … (laughs) … But! I support this project because it's clear! It has attitude. Although it lacks links to the surroundings of the site. But it is brave and I wish to give them a chance to develop their idea further, with more details.

B: It's fresh! This is probably a very young team. I love the garish colours! The densification strategy is unique and highly adequate. Their only mistake is that their urban design concept should address a bigger part of the Olympic Park – and that it should leave some public space around the stadium. This project is very energetic. It is a project with a different attitude. We should have enough courage to choose them as finalists for the second stage.

C: The proposition lacks connections with the river.

B: Well, they do address the river's banks: by not touching them!

D: Me, I would just throw this project away. It does not address our objectives and problems. There is too much focus and new buildings surrounding the stadium of the Olympic Park.

E: The concept is very strong. It raises questions, for example regarding the relationship between the built and non-built. The project redefines the relations between the Olympic Park, the golf course and the rest of the city. It's risky but it deserves to be examined in depth.

F: The strength of this project lies in its simplicity. It is not executable at this level, but it's a strong concept!

G: Well, I don't see anything positive in that proposition.

By means of Vignette 1 we enter a discussion that took place in the first of three rounds of elimination. As proposed by the professional advisor, during the first round of elimination, each entry was introduced by one of the jurors working as an architect (for one to two minutes) in order to assist the non-architect jurors. Each project was then examined and discussed by all the jurors for a period of four to six minutes. After that collective discussion, the president of the jury proceeded to a vote by show of hands. If the proposition received more than half of the votes (six jurors), it was admitted for the second round of elimination. However, the professional advisor clearly explained that it was always possible, until the end of the judgement procedure, to bring back an eliminated project in order to discuss it again. As he put it: 'It's never too late. Competitions are a democratic process and nobody should leave here with any regrets. Everyone is free to defend, at any moment, a project that seems interesting to him.'

The discussion presented in Vignette 1 concerns one of the fifty-two projects submitted. Since it is the first round, it is the first time that all the jurors have collectively examined and discussed this particular project.

The assigned architect (juror A) introduces the project and opens the discussion by sharing his personal observations and judgements. Juror A is clearly supporting the proposition: he highlights the originality and the strong concept of the project, even though he admits that some aspects left unresolved would need to be reworked for the second stage of the competition. He wishes to 'give them a chance' (line 4) to develop their 'brave' (line 4) proposition. The president of the jury (juror B) agrees and supports his opinion. Juror B formulates an interesting speculation to explain both the originality and the clumsy aspects of the project by suggesting that 'a very young team' (line 7) probably designed the project. He then appeals to the 'courage' (line 11) of the judges to support a project 'with a different attitude' (line 11). The discussion then sheers off with the intervention of the representative of the client (juror D) as he clearly indicates his opposition to the project. He argues that personally, he 'would just throw this project away' (line 18). Juror D does not mention any specific design aspects to explain his position other than 'there is too much focus and new buildings surrounding the stadium of the Olympic Park' (line 19–20). After two moderate comments from two technical experts, the local architect (juror G) formulates the last comment of the discussion. The opinion of juror G is clearly against the proposition as he simply indicates that he does not 'see anything positive in that proposition' (line 30).

During this short discussion, two divergent judgements regarding the proposition are clearly enunciated and the elimination vote reflected that division, as five jurors voted in favour of keeping the project for the second round of elimination while six voted against. Therefore, the project was eliminated.

Vignette 2

B: This project proposes a unique strategy.

A: This concept is completely new. We should really give the team a chance to develop their idea. Since we have three projects that are alike, let us please exchange one of them for this one.

D: Again, let me repeat, this idea is not what we asked for.

B: This team read the brief but they responded in a completely different way. The proposal plays with a clear juxtaposition of the dense built area – this garish rectangle – and the large green natural spaces. It's a new way, a different approach to connect the site of the Olympic Park to the landscape.

H: Okay, but then what kind of comments should we make to the team if we decided to keep this project for the second stage?

B: They would have to show us, for example, how their strategy could be extended to the overall Olympic Park's site, as well as the connections that need to be developed between the different areas of the Park and between the Park and the surrounding city quarters.

H: And what about the sustainability?

[Silence – the question remains unanswered]

I: We could also say 'strong vision, bad design'.

J: I do not understand that project. People here do not want density. They want space. They do not want their neighbourhood gentrified. Why should we select a project for the second stage when we already know very well that it will be eliminated anyway?

B: A jury meeting is a democratic process. Each member is free to express his opinion.

J: This would be a waste of time – and money.

B: It is crucial that we select different visions for the second stage.

J: Again, a project with so little chance of winning must be eliminated right now.

K: Anyway, is that really a team that we would like to work with? I don't think so. This is obviously a young and inexperienced team.

E: According to the brief and the title of the competition, the vision sought is that of a 'metamorphosis' of the whole site – and some of the projects selected as finalists so far do not even have a vision. It would be nice to have at least one finalist project that has a very strong vision.

By means of Vignette 2 we enter a discussion that takes place in the third and last round of elimination of the competition. In that last round, eleven projects are still in the race. The jurors therefore have to select five to seven of them (or eliminate four to six of them) as finalists for the second stage of the competition. For that last round, jurors have about forty-five minutes left to make their decision. Pressure and stress rise among the jurors. The president of the jury effectively animates the discussion while

Jury boards as 'risk managers'   125

the professional advisor strictly controls the remaining time. The first five projects are examined in about seven or eight minutes; very few comments are made. From those five projects, three are selected and two are eliminated.

The vignette starts with juror A interrupting the discussion. He notices that the overall approaches of the three projects selected so far are very similar. He suggests eliminating one of them and replacing it with the project that was labelled as 'garish' in the first vignette – although it was eliminated during the first round. As some jurors appear surprised by that proposition, the professional advisor reminds that, as he said at the opening of the judgement procedure: 'any juror is allowed to bring back any project at any moment'. Two of his assistants bring the boards of the 'garish project' back to the jurors.

At this moment, the discussion presented in second vignette is taking place. The judgements and opinions are as divergent as they were in the first discussion: some jurors are impressed by the originality of the design and are almost begging the jury to give it a chance for the second stage, while others strongly dislike the project and are firmly opposed to the idea of 'giving them a chance' (line 3). However, the enthusiasm of the jurors in favour is not convincing enough since the new vote leads to the same result as in the first round: five jurors vote in favour of selecting the project as a finalist while six vote against.

The interest of that discussion resides in the different natures of the arguments enunciated. On one hand, the jurors in favour underline the 'unique strategy' (line 1) of the project, its 'completely new' (line 3) concept and that compared to the other submissions, the team 'responded in a completely different way' (line 9) to the brief. Regarding the design strategies, they point out the quality of the project's 'clear juxtaposition of the dense built area [...] and the large green natural spaces' (lines 10–11) and 'Its [...] new approach to connect' (line 11) the different parts of the site. The jurors in favour also propose recommendations that would help the team improve their proposition for the second stage. Although they underline its potential, the jurors are aware of the weakness of the project and they qualify it as a 'strong vision, bad design' (line 26). Once again, they plead that the jury should 'give [them] a chance' (line 3) since there are already 'three projects that are alike' (line 4) among the finalists, and that it is 'crucial that we select different visions for the second stage' (line 36).

On the other hand, the jurors against the 'garish project' argue that 'this idea is not what we asked for' (line 7) in the brief. They question the pertinence of 'giving them a chance' and ask 'what kind of comments should we make?' (line 14) in the recommendation for the second stage. They also question the sustainability (line 22) approach proposed by the team. Here, it is important to underline that although these questions do not clearly reflect a position against the project, the in situ observation allows us to affirm that considering the attitude and the tone of the jurors, these were rhetorical questions expressing disapproval. The jurors against also argue that future users 'do not want density' (line 28) and 'gentrified' (line 29) architecture; 'they want space' (line 28).[5] They consider that 'a project with so little chance of winning must be eliminated right now' (line 38). Finally, interpreting the unconventional graphic design preferred by the team, the jurors against speculate about the age and the experience of the team and formulate the following argument: 'Anyway, is that really a team we would like to work with? I don't think so. This is obviously a young and inexperienced team' (lines 40–41).

Comparative exercise

Through the vignettes, we can observe that the jurors defending the proposition underline the originality of the concept by identifying the innovative design aspects of the project through concrete examples from the design as presented in the boards. They speculate on the conceptual potential of the project despite the various weaknesses. On the other hand, the jurors against the selection of the proposition consider that the project does not reach the brief's objectives. To support their position, they build their argumentation on the supposition that the team is too young to carry such a commission. They also speculate, but this time on the team's lack of experience, even though the competition's procedure was completely anonymous.

However, comparing the two sets of arguments reveals a major difference in the way those speculations are built. Although both the jurors in favour and those against perceive the 'garish project' as *risky* and both formulate speculations in order to build their judgement and convince the rest of the jury, the arguments used are distinct. The jurors in favour speculate about the design potentialities of the concept and the ability of the team to resolve them through a second phase of design. They perceive the risks as low since selecting them as finalists does not mean a contract or engagement: the project could always be eliminated later in the competition's procedure (during the second phase's judgement). They evaluate that it is worth taking the risk to select the team as a finalist since this risk is perceived as a potential good investment. In their opinion, it could represent a great benefit for the competition. Indeed, the jurors are confident that the proposition would finally appear to be a good and complete project after a second phase of design. In contrast, the jurors against the project speculate about the age of the team and their potential lack of experience. They evaluate that the risks are too high and not worth taking. In their opinion, selecting this team for the second phase would be a waste of time and money. They speculate and argue that the team is too young and inexperienced for such a commission and in order to defend their judgement, the main argument raised concerns the financial risk.

As we can see, two types of speculation were formulated. On the one hand, jurors in favour *presumed*. On the other hand, jurors against *assumed*. The jurors in favour presumed that the team could fix the identified weaknesses. They believed the project had potential and asked the jurors to 'give them a chance' to improve their proposition. However, isn't the word 'chance' acknowledging the possibility that the team could also fail to deliver a better project in the second phase? On the other hand, the jurors against assumed the lack of experience of the team. They attributed the originality and the unconventional graphical presentation preferred by the team to young architects. But was their speculation correct or incorrect, considering the anonymous procedure? That difference might be very subtle but appears to be fundamental in order to understand how risks are managed in the context of architectural competition juries.

As illustrated by the selected vignettes and as observed through the in situ observation of various architectural competition juries at work, jurors regularly consider the experience – or the lack of experience – of a firm in order to build their argument or judgement in favour of or against the selection of a project as finalist or laureate. However, the consideration of that parameter, or criterion, for the judgement of the quality of a project generally leads to divergent opinions: some jurors are reluctant to nominate young architects while others are willing to nominate them. Somehow, it seems that for some jurors, the selection of a young firm represents a higher risk for the completion of the project than for other jurors. Therefore, the analysis of the vignettes leads us to formulate, through risk management theories, a hypothesis regarding how architectural competition jurors deal with risks when judging architectural quality.

Risk and collective responsibility

The analysis of the vignettes reveals that in an architectural competition jury, the jurors might speculate on the experience of the team and discuss the consequences it would represent to choose a young team as finalist.[6] But, considering the presented example, why are the jurors speculating on the author when 1) the experience of the team is not mentioned in the judgement criteria or in the application conditions (except that all participants had to be architects); and 2) the competition deliberately follows an anonymous procedure, which means that the project should not be evaluated regarding its authors but only regarding the quality of the submitted proposition?

As underlined by the jurors against, the selection of the 'garish project' represents an investment of money that might lead to nothing. Of course the jurors in favour perceived it otherwise. But the question is, since it is already budgeted – and as in both stages of the competition, only one finalist will win while the others won't – why would some jurors be preoccupied by the competition's expenses? In fact, as observed in situ, although the money was available and budgeted for five to seven finalists, each paid about 50,000 Canadian dollars for the second stage of the competition, only five were selected. Furthermore, the jurors that eliminated the 'garish project' finally insisted on selecting only five finalists for the first stage of the competition since it would represent 'an economy of 100,000 Canadian dollars'![7]

In an architectural competition, the task of the jury is to judge the quality of the proposition regarding a specific set of criteria in order to select the best project. In this sense, why would a jury be preoccupied by aspects, such as the budget, that they are not responsible for? Why would a jury try to anticipate or predict a 'future' – the competence of the team – that was intentionally ignored by the anonymous procedure chosen by the competition's organizers? More generally, why did some jurors feel, to a certain degree, personally responsible for the competition's output? While managing the potential risks related to a project in order to build the architectural and qualitative judgement about the submitted projects, jurors also appear to be preoccupied not only by the architectural aspects of the future realization of the project, but also by the 'moral' consequences their judgement and their decisions could represent. In risk management theories, this phenomenon is identified as the 'intellectual risk':

> As well as physical risk and financial risk there is a third type: intellectual risk. A person takes on intellectual risk when he sets out to provide an adequate explanation for something where previous attempts have failed, and he takes an intellectual risk when he sets out to question the validity of some explanation which most people believe to be perfectly adequate. In taking an intellectual risk a person stands to lose neither his life, nor his fortune, but his credibility. Since knowledge, like air travel, is usually believed to be so useful there are strong disincentives to intellectual risk taking, and anyone who wishes to take such risks would be well advised to immerse himself in some relatively useless area of knowledge, such as anthropology.
>
> (Thompson, 1980. See also Younés, 2012)

In other words, the jurors of an architectural competition also appear to be preoccupied by the impact of their decision, not only in architectural terms, but also as a potential risk for their own credibility. In fact, financial preoccupations – 'this project seems too expensive' – are regularly raised by the client's representative – frequently an elected political figure – as an argument in favour of or against a project. Furthermore, these comments are, most of the time, followed by another comment, which somehow reflects the preoccupation of the juror to please his or her electorate, such as 'the public, the tax payers would not like it' or 'if we choose this proposition, it will be very difficult or even impossible to convince the public that it's a good project. It's too 'avant-garde.'[8]

Therefore, the hypothesis formulated through the lens of risk management theories would be that competitions are not only *judging devices*, they are also *judged processes*. The judgement can be public, journalistic, historic, architectural … it does not matter. The point is that some jurors feel affected by the intellectual risk their decision making process represents. In fact, jurors seem to be sensible enough to that 'second layer' of judgement that, in addition to acting as *judging agents*, they also think as *judged agents*. However, not all jurors perceive risks in general and intellectual risk in particular the same way. This would be one of the reasons why some jurors are more likely to formulate presumptions while others are more likely to formulate assumptions in order to convince and to build their argumentation.

Finally, these 'extra layers' of judgement might be so important that it could explain why, in the official jury report of all the cases of architectural competition juries we observed on the field, arguments such as the speculated experience of the team are never reported. Indeed, the jury reports appear to systematically omit the most controversial speculations formulated when listing the arguments, motives or reasons that lead to the elimination of the selection of any of the propositions and that participated in the construction of the collective judgement of the projects.

Conclusion

How can the concept of risk and risk management theories inform the analysis of decision making within jury boards of architectural competitions? What are the risks discussed in jury sessions? How are they perceived and judged? Why are certain parameters considered risky by some jurors and safe by others?

The risks associated with the realization of an architectural project cannot be predicted (other market products, scientific experiments, etc.), since they are always, somehow, built like prototypes. However, in a competition jury, the actors try to anticipate the risks associated with each project in order to evaluate how the project (and/or its authors) will perform in the future. However, since the risks they might carry or be exposed to are extremely difficult to measure and to anticipate, speculation often occurs during an architectural competition judgement process. The vignettes analysed for the purpose of this chapter show that two types of speculation can be formulated in order to address the perceived risks: assumptions or presumptions. Therefore, some jurors seem to be preoccupied by the 'behaviour' of the project in the future, qualities as they are presented in the boards, while others are also preoccupied by the client's interests or the intellectual risk associated with the decision they are making – how they will be judged after the public reception of the results.

In summary, what risk management theories reveal when linked to the analyses of architectural competition juries is:

1. that architectural judgement includes more aspects than the intrinsic qualities of the projects as they appear in the submissions (boards and texts) – many other aspects are taken into account, even though some of them might be the result of pure speculation. The speculations observed in the presented vignettes about the age and the experience of the team are relevant examples.
2. that architectural judgement in the context of architectural competition juries is much broader than the jury meeting period. The jurors of a competition are not the exclusive 'judging agents'; the procedure, the choice of the laureate, the outcome of the competition, and so on might be subject to the critics. The press, the critics and the public also judge the decisions of the jury when the results are released – and jurors seem to be, to different degrees, conscious of it, and take this reality into account while judging.

However, the analysis proposed through this chapter reveals that speculations are not done at random or by hazard, they are necessary and constitutive of architectural judgement in the context of a competition's jury. The Opéra de la Bastille case remains a great example for the problematic tackled in this research since the jury based its decision in part on the speculation that the author of the project was the renowned architect Richard Meier. However we could question or remain critical regarding the nature of the speculation: whether they are assumptions or presumptions can make a great difference in the way the collective judgement is built:

> The primary problem with risk assessment is that the information on which decisions must be made is usually inadequate. Because the decision cannot wait, the gaps in information must be bridged by inference and belief, and these cannot be evaluated in the same way as facts. Improving the quality and comprehensiveness of knowledge is by far the most effective way to improve risk assessment.
> (Adams, 1995)

> Ulrich Beck interprets risk as brought into being by social entities, such as science, law, politics, and the mass media, which define, select, and manage risks. Beck argues for the social construction of real and imagined risks. Since risk implies an uncertain future, real dangers as well as concerns, fears, or imagination are indissolubly parts of risk. Risks are always real and constructed.
> (Zinn, 2008)

Notes

1 These comments are fictional examples based on the data collected through our various fieldwork observations.
2 The only exception concerns the competitions seeking approval by the Royal Architectural Institute of Canada (RAIC – Architecture Canada, www.raic.org), the architect's Canadian association. To be approved by the RAIC, competitions must follow an anonymous procedure. However, only a few Canadian public buildings' competitions decide or need to be approved by the RAIC in order to be organized, so mandatory anonymous procedures remain exceptional.

3 The programme and the composition of the jury of the competition have been modified in order to respect the strict policies of confidentiality applying to any person participating in or observing an architectural competition jury's procedure. The original competition took place in the 2010s in Canada.
4 More precisely: A, B, D, G, H, J and K are jurors, that is, they are allowed to vote. C, E and F are technical experts consulting the jury, that is, they are not allowed to vote. A and B are both foreign architects. B is president of the jury. D is a civil engineer and he represents the client's interests. G is a local architect. I is the professional advisor of the competition.
5 In our opinion, this comment needs to be taken carefully since all the submitted projects of the competition proposed densification and gentrification of some sort, since these were two of the main objectives of the competition ('densification' and 'revitalization'). As another juror underlined right after this discussion: 'according to the brief and the title of the competition, the vision sought is that of a "metamorphosis" of the whole site.'
6 During our fieldwork, this phenomenon was observed in more than one case.
7 The numbers were modified in order to respect the strict policies of confidentiality.
8 These comments are fictional examples based on the data collected through our various fieldwork observations.

References

Adams, J. (1995). *Risk*. London: UCL Press.
Boutinet, J.-P. (1990). *Anthropologie du projet*. Presses Universitaires de France.
Chaslin, François (1985). *Les Paris de François Mitterrand: Histoire des grands projets architecturaux*. Paris: Gallimard.
Chupin, J.-P. (2011). Judgment by design: towards a model for studying and improving the competition process in architecture and urban design. *The Scandinavian Journal of Management*, 27(1), pp. 173–184.
Silberberger, J. (2011). *A Qualitative Investigation into Decision-Making within Jury Boards of Architectural Competitions*. Fribourg: UniPrint.
Silberberger, J. (2012). Jury sessions as non-trivial machines: a procedural analysis. *Journal of Design Research*, 10(4), pp. 258–268.
Söderström, O. (2000). *Des images pour agir. Le visuel en urbanisme*. Lausanne: Éditions Payot.
Sudjic, D. (2006). Competitions: The Pitfalls and the Potential. In C. Malmberg (Ed.), *The Politics of Design: Competitions for Public Projects*. Princeton: The Trustees of Princeton University.
Thompson, M. (1980). Aesthetics of Risk. In R. C. Schwing and W. Albers (Eds), *Societal Risk Assessment: How Safe is Safe Enough?* New York, London: Premium Press.
Weick, K. E., Sutcliffe, K. M., and Obstfeld, D. (2005). Organizing and the process of sensemaking. *Organization Science*, 16, pp. 409–421.
Younés, S. (2012). *The Imperfect City: On Architectural Judgment*. Aldershot: Ashgate.
Zinn, J. O. (2008). *Social Theories of Risk and Uncertainty*. London: Blackwell Publishers.

8 Competitions beyond spatial specifications

An interview with Dietmar Eberle

The second part of this book has elaborated on decision making within the jury boards of architectural competitions. It has shown how judgement criteria evolve over the course of jury deliberations by means of an interaction between jurors, architectural principles or values and the entries submitted. The three preceding chapters have furthermore illustrated that detailed specifications provided in the brief do not play as prominent a role as one would assume. The following interview complements this finding. Dietmar Eberle criticizes the current trend within the competition business to approach complex building tasks by defining a large number of detailed spatial specifications. On the contrary, Eberle suggests turning around this process of formalization by arguing for procedures that courageously work against the idea that the winning entry should be the one that responds best to the spatial specifications requested. The role of the competition – according to Eberle – should be understood as a way to develop an effective and durable client–architect relationship while engaging with a given design problem.

An interview with Dietmar Eberle, founding partner of be – Baumschlager Eberle

Baumschlager Eberle, which has offices in Lustenau, Vaduz, Vienna, St. Gallen, Zurich, Hong Kong, Berlin, Hanoi, Paris and Hamburg, is a highly successful participant in architectural competitions. Eberle has procured a number of projects by taking part in architectural competitions. Projects include: the mixed use project 'Solids IJburg' in Amsterdam, the Netherlands; the 'Maison du Savoir' that is the main building of the University of Luxembourg in Belval; the Zurich residential towers 'The Metropolitans'; as well as the BNP Paribas Fortis Bank Headquarters in Brussels, Belgium, which is scheduled for completion by the end of 2021. In addition to participating in competitions, Eberle also frequently takes the role of jury member in a number of international competitions.

Eberle studied at Vienna University of Technology in Austria. He co-founded the Baukünstler design movement in Vorarlberg, Austria in 1979. From 1985 to 2010 he worked with Carlo Baumschlager, and since 1983, he has taught in Hannover, Vienna, Linz, Zurich, New York and Darmstadt. Currently, he is a Professor of Architecture and Design at the Swiss Federal Institute of Technology in Zurich, Switzerland and has led the ETH Wohnforum – ETH CASE (Centre for Research on Architecture, Society and the Built Environment) since 1999.

An interview with Dietmar Eberle

Dietmar Eberle (DE) was interviewed by Ignaz Strebel (IS) and Jan Silberberger (JS).

IS: How important were architectural competitions for you at the beginning of your career?

DE: For me, architectural competitions were not important at all. I did not take part in any in the first ten years of my professional life – simply due to the fact that my values did not correspond to the values of the initiators of architectural competitions. Regarding the office, the projects that we built in the beginning were all direct commissions. Today, this has changed. We take part in competitions, however, not in open competitions because too much randomness comes with them. We take part in competitions by invitation and we take part in procedures involving a pre-qualification. Regarding the latter, we decide on the basis of the challenge that the task poses and the size of the project.

JS: What role does the jury composition play in this context?

DE: Undoubtedly, the composition of the right jury is the most crucial task for the building contractor. The composition of the jury directly relates to the quality of the competition's outcome. The jury has a larger effect on the quality of the result than the competition programme or the participating offices. But when deciding whether to participate, the composition of the jury does not play a role. I have this principle. If we do competitions, then we do what we believe in; what we think is right. And if the jury or the client thinks the same, then we are pleased. If not, we are still pleased. Then we just have a different opinion. That's all. This strategy has not worked quite so bad after all. Based on the competition programme you can recognize the values of the awarding authority. So if the competition programme tells you that you cannot get to common values, then you don't participate. What's more decisive though is the type of procedure. Once again, we do not take part in open competitions.

IS: So of all the competitions that you have won not a single one was open?

DE: You have to keep this in perspective. Architectural quality does not emerge that way. If you want a certain quality, then you must not run an open competition. That's in fact rather logical. Highly qualified offices have other possibilities and they don't like this kind of procedure that involves a great deal of randomness for the participating office. This sounds very exclusive, but it is true that nowadays, open competitions produce rather mediocre results. If you want to get around this, that is, if you want to enhance the quality of the results of competitions, then you do not even have to invent something new. You only have to recognize the fact that there is a whole range of procedures to choose from and that the most decisive task is to choose an adequate one. Different building tasks require different procedures. It's as simple as that. So you have to categorize which questions you want to pose, and then which procedures are suited for producing answers to these questions.

IS: How would you proceed then for a XL building task such as a large hospital?

DE: One has to see that in the end nothing can replace this essentially important dialogue process between the client and the executing architecture office. In this respect, architectural competitions should not focus on the organization of the internal spatial specifications of the planned building, which they unfortunately still do, especially in Switzerland. Rather, they should focus on the contribution of the building to the public space, on its material consistency as well as on the rigour of the methods and principles that are applied. The more complicated a building, the less suitable the open competition – and particularly not in the form as it is usually executed nowadays, which puts the focus on spatial specifications. This is a method of the nineteenth century.

JS: Many architects today complain about extremely overloaded competition programmes. For example, the specifications in regard to fire protection become increasingly extensive.

DE: It's always society that defines the contemporary questions. Each period has its own questions. In the early 1960s there was central heating. Each competition entry had been equipped with a scheme of central heating. In the 80s one could experience that the focus turned to the exterior spaces. The 90s then were simply a question of energy efficiency. Of every competition entry an insane documentation on this topic was demanded. In the last ten years fire protection was the driving force, since the regulations on fire protection have vastly changed throughout Europe. At the moment there are two trends of new issues that are becoming increasingly important. One is the role of the user, that is, how people relate to architecture. The other is the use of three-dimensional calculation methods like Building Information Modelling. In between we had a phase where at one time accessibility for disabled people was discussed. I always say it somewhat sarcastically. Every three years there is suddenly a new sensation that becomes super important, and for me that means nothing else than that there is lack of consciousness or experience in dealing with these issues at a time when they are most strongly demanded. So why do competition programmes propose solutions such as the 'Minergie'-standard and therefore underestimate that participants are able to produce new insights, new findings in regard to, for example, energy questions, which essentially shape design? So if you ask yourself why our publicly used infrastructure is so expensive and often so meaningless, this is exactly why. The public awarding authorities have a false understanding of competition. They simply ask the wrong questions, especially in regard to energy efficiency. Today we suffer from the immense amount of maintenance that modern buildings require, which is especially hard for the public sector. It is completely absurd. We equip our buildings with way too much technology – with a half-life of ten years – which involves very high maintenance costs.

JS: What kind of paradigm shift would that involve?

DE: The crucial question in the twenty-first century is to stop planning based on the idea of 'designated use' because history teaches us that this changes rather rapidly and has a maximum life span of one generation. Who wants to work as we did twenty-five years ago? Who wants to live like their parents? Who wants a hospital run according to the standards from twenty-five years ago? So the essential question that has to be dealt with is not the building's designated use, but its contribution to the public space. This is what determines the building's value in the long run. The most important user of the building is not the one who enters it and uses its inside, but the passer-by. I think that the passer-by will become the most important user of architecture in the future.

IS: How many competitions are organized out there reflecting such a way of thinking?

DE: In Switzerland, the state of the discussion is still somewhat modest in intellectual terms, but take the example of 'Solids IJburg'. Do you know what kind of competition that was? This was a selection process based on a discussion of values. They invited ten architecture offices, but there was no project to be submitted. Or take another example, the hospital which I have built in Kortrijk, Belgium. This has 1,400 beds. That was the same procedure. We discussed values without a project. There was not a single drawing. So this is a matter of the values that you have. They gave you ten questions and said, 'Tell us what do you think about that'. I just want to say that there are an incredible number of different selection methods, which by the way are all competitions according to EU regulation. 'Solids IJburg' was a competition, but the way it was carried out completely differed from how competitions are conducted here in Switzerland. There were absolutely no spatial specifications! There was nothing. There was just a formulation of the problem and there were various questions. They just asked you what you think about certain questions that ultimately relate to building and the construction of buildings. In the case of the hospital in Kortrijk, all the doctors were there. All these people knew one crucial thing. Normally an architectural project takes several years. So they need an architect on whom they can rely. They need people that they can be sure will stay consistent. People that they know are open to dialogue, and who are competent to develop solutions to arising problems. Then the client said, and it is exactly the client's formulation, 'it would be nice, if we could work together and develop something'. Then you start to work from the big to the small, from the public to the private. In the end it is not important to ask yourself how to organize an operating room. There are hundreds of specialized planners and thousands of companies, which know exactly how to do that. There is so much objective knowledge in regard to this issue that you only have to say which model of operating room you pick. Indeed, this reflects that the competition system emerged within public procurement and is impoverished by a simple fulfilment of spatial specifications.

IS: But how do you approach this understanding as a juror?

DE: First of all, I want to say that I agree to be a juror for personal reasons. So either I work for someone with whom I have a long personal relationship or the task posed interests me. Good jury work first of all engages in understanding the competition entries presented. This is a comparative process between the presented entries. I would say that only bad, stupid, mediocre or uneducated judges compare competition entries with what they have in mind. Jury work is not about finding some idealistic, ideal-typical solution to a problem. It is about a much simpler question of which one of the presented competition entries is the most effective, the best, the most sustainable one in relation to the task posed. That means I have to engage extremely accurately with what I have in front of me.

JS: Research on architectural competitions distinguishes between 'evaluation', which is about the quantitative assessment of individual aspects of competition entries, and 'judgement', which is the qualitative, holistic interpretation of the entries. It is then argued that juries, especially, for example, since the introduction of sustainability standards such as LEED, increasingly 'evaluate' and that the qualitative judgement is more and more superimposed by these evaluations.

DE: Of course there are evaluations that represent quantitative methods. In fact, they relate not just to energy standards, but also for instance to the management of spaces. There are many criteria that are assessed quantitatively. And then, obviously, there is the qualitative interpretation of architectural design. If you call that 'judgement', fine. If one says that evaluations replace judgements then this is a synonym for poor jury work. I can only repeat that the quality of the jury plays the pivotal role for the quality of the competition's outcome.

IS: Such discussions then should therefore take place prior to the jury sessions?

DE: Indeed, that's why there is a distinction in Europe. First you have the preliminary examination that performs the evaluations. People that are not part of the jury board usually do such pre-screenings. However if you then have people in the jury who are too involved in these pre-screenings, then that means only one thing. They are overstrained in their task of judging.

Part III
Making the built environment

9 The obligatory passage point

Jan Silberberger and Ignaz Strebel

Introduction

As we have shown in the Introduction to this book, research on architectural competitions can be considered a well-established field nowadays. However, there is still relatively little knowledge on the direct follow-up of the competition procedure, that is, when (representatives of) the client and the executing architecture office are faced with the task of transforming the winning entry (which is a proposal for a solution) into a physical building (an actual solution).

In this chapter we intend to do something unorthodox: to say that competitions are not finished when the results are announced. We will elaborate on how competitions act on the follow-up of the building process. To start with, it is useful to take a brief look at the technical terminology developed in the field of project management in construction above ground. Muhm (2014) assigns the competition procedure to a phase called 'Development of a concept' and names the subsequent phase 'Planning'. According to Menz (2009), the building process can be divided into six phases: Strategic Planning (phase 1), Preliminary Studies (2), Project (3), Invitation to Tender (4), Implementation (5) and Management (6). Using this latter terminology, the chapter at hand focuses on the transition between preliminary studies, to which the competition ('selection procedure') is assigned, and project (see Table 9.1).

In his definition, Menz (2009, pp. 207–211) refers to the 'Service Model' as provided by the Swiss Society of Engineers and Architects (SIA), the so-called 'Regulation SIA 112' (2001). This regulation coordinates the client's responsibilities and the tasks of the various planners involved during the building process by defining subphases and goals of subphases. As Table 9.1 shows, during the project phase, the client and the executing architecture office primarily deal with economic aspects: they focus on optimizing the project's concept, costs and profitability. That is, while the architectural competition is primarily conducive to architectural as well as urban design aspects of quality, the project phase puts aspects of cost effectiveness at the centre. This shift, we presume, almost necessarily results in a reformulation of the winning competition entry during the project phase.

The chapter at hand precisely focuses on such reformulations. First, it will provide a detailed account of the reformulation of one particular winning competition entry during project. Then, second, the concept of translation as coined by Callon (1986) will be applied to provide an analysis, which also incorporates findings from another thirteen case studies (see Table 9.2). In this way, we aim at conceptualizing

Table 9.1 Phases, subphases and goals of subphases of the building process.

Phases	Subphases	Goals
Strategic Planning	Formulation of needs, solution strategies	Needs, goals and general conditions defined, strategy for solution determined
Preliminary Studies	Definition of the project, feasibility study	Procedure and organization defined, project basis defined, feasibility demonstrated
	Selection procedure	Project selected which will best meet the requirements
Project	Preliminary project	Concept and profitability optimized
	Construction project	Project and cost optimized, schedule defined
	Permit-obtaining procedure	Project approved, cost and schedule verified, construction credit granted
Invitation to tender	Invitation to tender, comparison of quotations, application for contract to be awarded	Contract ready for awarding
Implementation	Construction planning	Project ready for implementation
	Implementation	Building structure constructed according to specifications and contract
	Commissioning, completion	Building structure accepted and commissioned, final cost settlement accepted, defects corrected
Management	Operation	Operation ensured and optimized
	Maintenance	Fitness for use and value of the building structure maintained for defined period of time

the architectural competition as an 'obligatory passage point' (Callon, 1986) and the winning entry as a temporarily stabilized entity, which is threatened to be destabilized (again) during project. In doing so, we intend to shed light on the impact of competitions on the subsequent phases of the building process.

Methodology and selection of cases

The findings presented in this chapter are based on expert interviews with the clients as well as the winning architecture offices of thirteen housing competitions in Switzerland. As Table 9.2 shows, the sample selected for this study covers a large range of competitions with regard to the type of client, the 'scale' and type of the task, the location of the construction project as well as the size and standing of the winning architecture office. Yet, we are far from claiming that our data sample is representative. Rather, we would like to highlight the exploratory character of our study, and therefore work with an expressive sample.

Table 9.2 Basic facts of the construction projects investigated.

Title of project	Task posed	Client	Location	Form of procurement	Type of procedure	Winner
Neubau Wohn- und Gewerbesiedlung Kalkbreite	New construction of a housing (88 flats) and trade estate	Housing cooperative 'Kalkbreite'	Zurich	Project competition	Open	Müller Sigrist Architekten AG (Zurich)
Rosengartenhof Küssnacht	New construction of 12 apartments and (ground floor) commercial space		Küssnacht am Rigi	Study commission	Invitation	Roman Hutter Architektur GmbH (Lucerne)
Münster-Vorstadt Sursee	New construction of apartments for the elderly	City of Sursee / Azor AG	Sursee	Ideas competition	Invitation	Roman Hutter Architektur GmbH (Lucerne)
Mehrgenerationenhaus Winterthur	New construction of an apartment complex (150 flats)	Housing cooperative 'GESEWO'	Winterthur	Project competition	Restricted	Dachtler Partner AG (Zurich)
Wohnüberbauung Landolt-Areal, Zürich-Enge	New construction of an apartment complex (52 luxury flats)	Agruna AG	Zurich	Study commission	Invitation	Bünzli & Courvoisier Architekten AG (Zurich)
Ein Haus für junge Menschen	New construction of a house for apprentices (12 rooms and a restaurant)	City of Zug	Zug	Project competition	Open	Lando Rossmaier Architekten AG (Zurich)
Wohnüberbauung Gries, Volketswil	New construction of a housing estate (GFA: 15,000 m2)	Allreal AG	Volketswil	Study commission	Invitation	hbf Architekten (Zurich)
Pilotprojekt Wohnhaus Aescherstrasse 10/12	Construction of a pilot project for sustainable housing (7 flats)	City of Basel	Basel	Study commission	Restricted	Osolin & Plüss Architekten + quade architects (Basel)

(*continued*)

Table 9.2 Continued.

Title of project	Task posed	Client	Location	Form of procurement	Type of procedure	Winner
Wohnsiedlung Obsthalde in Zürich-Affoltern	New construction (replacement) of a housing complex (NIA: 5,000 m2)	Housing cooperative 'Süd-Ost'	Zurich	Study commission	Invitation	EMI Architekten AG (Zurich)
Ersatzneubauten Wohnsiedlung Himmelrich 3, Luzern	New construction (replacement) of a housing complex (240 flats) + commercial space	Housing cooperative 'abl'	Lucerne	Project competition	Restricted	Enzmann Fischer AG (Zurich)
Umbau der ehemaligen Papiermühle	Reorganization / conversion of a former mill into a residential building and office space	City of St. Gallen	St. Gallen	Study commission	Restricted	Bischof Gruber Architekten (Zurich)
Ersatzneubau Wohnsiedlung Letzigraben	New construction of a housing complex (95 flats)	Housing cooperative 'eigengrund'	Zurich	Project competition	Restricted	Von Ballmoos Krucker Architekten (Zurich)
Ersatzneubauten Siedlung Buchegg 1+2	New construction (replacement) of a housing complex (113 flats)	Housing cooperative 'Waidberg'	Zurich	Project competition	Invitation	DUPLEX Architekten (Zurich)

The obligatory passage point 143

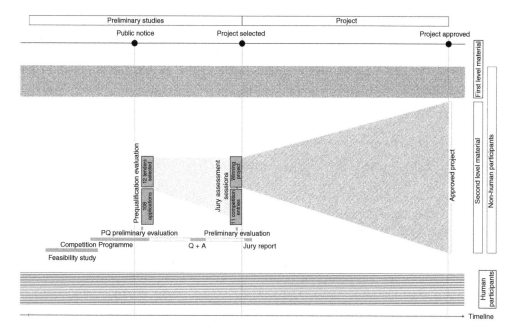

Figure 9.1 Project diagram used in interviews.
Source: Authors.

In the course of the interviews we provided our interviewees with a 'project diagram' (see Figure 9.1), which we invented on the basis of visualizations of procedural aspects of construction projects as have been introduced, for example by Rem Koolhaas and Bruce Mau in *S, M, L, XL* (1995) or Clare Melhuish in *Luis Vidal + Architects – From Process to Results* (2013), as well as on the basis of Muhm's (2014) graphs dealing with the (client's) possibilities to influence or alter a building's functions and form throughout the various phases of the building process. Our project diagram shows a vertical division into two main phases (preliminary studies, project) and a horizontal division into 'actors': human participants and non-human participants that are themselves divided into first and second level material. First level material comprises physical conditions (such as for instance the geology of the building site, the existing infrastructure or the existing building in the case of an extension). Second level material consists of all documents (plans, schemes, concepts, expertise, reports) and models produced during both depicted phases. The basic project stages (feasibility study, competition programme, preliminary evaluation, Q&A, jury assessment sessions, jury report, winning project and approved project) are already entered and decisive points in time – public notice, project selected, project approved – are highlighted with black dots. The representation of second level material within the project phase is an empty funnel, which linearly extrapolates the various design aspects of the winning competition entry (architecture, urban planning, structural planning, energy efficiency, cost planning, heating/ventilation/sanitation/electrical installations) in the same spirit as Melhuish (2013) suggests. The idea behind our project diagram was to illustrate and trace the connections, associations and relations between the different project stages and actors. We actively encouraged our interview partners to edit the project diagram and also to draw into the plans representing their competition entry (which we put next to the project diagram) in order to illustrate and support their statements.

In what follows, we will take a closer look at the first case listed in Table 9.2, the 'Kalkbreite' project competition. We will then analyse the proceedings within the course of this competition and its subsequent project phase in a way that also allows for integration of findings from the whole set of case studies listed in Table 9.2.

The Kalkbreite case – a brief history

As Table 9.2 shows, the Kalkbreite construction project, which was put into operation in July 2014, comprises eighty-eight flats (with a total net internal area of 7,700 square metres), communal spaces (with a total net internal area of 600 square metres) as well as commercial space (with a total net internal area of 4,900 square metres). The plot and its utilization, however, had been an issue of debate within Zurich's department of urban planning since the late 1970s (the Kalkbreite plot, which is owned by the city of Zurich, is heavily exposed to noise pollution and had until then solely been used as a parking ground for cars as well as for trams). Therefore, the awarding of the right to build, which was issued to the housing cooperative in autumn 2007, can be considered a break-through. Yet, along with this award came the obligation to conduct an architectural competition – supervised by Zurich's public building department. To prepare this competition, the housing cooperative and the city of Zurich commissioned a Zurich based architecture office to do a feasibility study, which was completed by the end of March 2008. On the basis of the findings of that study, the building committee of the housing cooperative together with staff from Zurich's public building department elaborated the competition programme for an open, anonymous project competition (the invitation to tender was issued on 18 July 2008, the jury's final decision on 27 February 2009). The competition attracted fifty-five entries and was won by Zurich based office Müller Sigrist Architeckten AG. Figure 9.2 presents a rendering, which was part of the plans Müller Sigrist Architekten AG submitted to the competition.

Only shortly after the architectural competition, that is, at the beginning of the project phase, the housing cooperative started to promote its commercial space to potential future tenants. At that time the cooperative envisaged an organic supermarket, a restaurant, several small shops as well as various small service providers, as a detailed list of spatial specifications in the competition programme proves. However, soon after the housing cooperative started negotiations with potential business tenants, a cinema exhibitor showed interest and agreed to lease space for twenty-five years guaranteed. That offer posed an incredible opportunity for the housing cooperative, since it not only meant economic security, but in addition could be considered to entail further tenants, such as for instance the envisaged café and bar. However, taking this incredible business opportunity meant that the winning entry had to be significantly modified, since a five-hall cinema, which had not been envisaged during the competition, had to be integrated into the existing structure of the building. The bearing structure had to be reworked, the staircases had to be moved and the floor plans of the flats located over the cinema had to be remade. Moreover, as Figure 9.3 shows, a two-storey 'oriel' had to be added since the cinema's floor space requirements could not be met within the given building depth.

Figure 9.2 Rendering representing the winning entry.
Source: Courtesy of Müller Sigrist Architekten AG and Raumgleiter GmbH.

Figure 9.3 Rendering representing the project that obtained the construction permit.
Source: Courtesy of Müller Sigrist Architekten AG and Raumgleiter GmbH.

The Kalkbreite architectural competition as an obligatory passage point

In what follows we will analyse how the Kalkbreite competition was used as a platform to integrate a variety of entities involved in the opening phases of the building process. To do so we will introduce and apply Callon's (1986) four moments of translation – problematization, interessement, enrolment and mobilization – which he elaborated when studying the re-cultivation of a bay in northwestern France with scallops as a controversy involving various human and non-human entities. When applying these rather abstract concepts to analyse the proceedings within the course of the Kalkbreite case we intend to produce an account that allows for integrating findings from the whole set of case studies we conducted (see Table 9.2).

Problematization according to Callon concerns the question of 'how to become indispensable' (1986, p. 204). In simple terms this means that an actor (which can be a human or non-human entity) defines a problem in a way that shows that he, she or it (i.e., the same actor) offers a solution for this problem to all other entities concerned. Interessement then refers to the question of how the other actors are 'locked into place' (1986, p. 207). Hence, interessement can be described as the attempt to win other entities over by detaching them from rival actors, that is, entities that try to involve them into their own enterprises. Enrolment as a third moment of translation concerns the definition, coordination and acceptance of roles. Enrolment describes successful interessement: the 'primum movens' (Callon, 1986, p. 204), that is, the main entity or actor has been able to convince the other actors to accept the roles it has designated to them. Finally, mobilization as a fourth moment of translation describes the various actions or displacements that have been necessary for one entity (the primum movens) to become accepted as spokesperson for the other entities.

At the outset of the Kalkbreite competition we can identify four main entities:

1. the city of Zurich represented by its department of urban planning and its public building department;
2. the (building committee of the) housing cooperative;
3. the set of potential future business tenants; and
4. the future inhabitants.

Each of these entities has a clearly defined goal: the future inhabitants aim at residing in the best possible way; the future business tenants seek to become profitable enterprises or strive to assure a long-term profit; the housing cooperative aspires to realizing an optimum building; and the city of Zurich aims at increasing its quality of living. When looking at the Kalkbreite case in retrospect, we would describe its primum movens as a composite entity combining representatives of the city of Zurich (its urban planning department as well as its public building department) and the housing cooperative.

Before applying Callon's moments of translation, we have to point out that, in contrast to the case studied by Callon (1986), where marine biologists, fishermen and scallops pursued conflicting goals, the participating institutions and individual actors in the Kalkbreite case have to be understood to not stand against each other in polarizing positions. Hence, a lot of translation work, so to speak, can be taken as a given. Already in the autumn of 2007 when the housing cooperative was awarded the right to build, a consensus existed as regards the development of the former parking

Table 9.3 Overview of the various displacements (= mobilization).

Initial entity	First translation	Second translation
Set of potential future business tenants	Shortlist of targeted business tenants as compiled by the housing cooperative	Four pages of spatial specifications forming an essential part of the brief
Set of potential future inhabitants	Group of 30–40 potential future inhabitants discussing needs and wishes	Nine pages of spatial specifications forming an essential part of the brief

lot. Furthermore, there was mutual trust between the four main entities listed above almost from the start of their interaction. Instead of rival entities with conflicting interests, the housing cooperative, Zurich's departments of urban planning and public building, the potential business tenants and future inhabitants, as well as the competing architecture offices, already saw themselves as partners pulling together. And, most importantly, since the Kalkbreite plot remains an asset of the city of Zurich – the city decided to lease it to the housing cooperative for sixty-two years – the housing cooperative as a lessee was legally bound to conduct an architectural competition supervised by the city's public building department. Therefore, in the Kalkbreite case, the competition procedure literally constituted an obligatory passage point on which all entities listed above had to rely.

Hence, instead of discussing successfully completed processes of problematization and interessement (and consequentially enrolment), we will look at how the set of potential future business tenants and the future inhabitants have been translated into entities, which the composite primum movens could control or speak for respectively (see Table 9.3).

In the case of the potential future business tenants, an idea of sought-after types of business had been translated into a shortlist precisely defining the targeted group. In a second step, by means of intensified discussions within a committee of the housing cooperation and by drawing on empirical values, this condensed list of targeted commercial tenants had been translated into a highly detailed list of spatial specifications that displays the demands of each type of use and assigns exact square metre figures. This extensive list then made its way into the competition brief. Featured on four pages it served as an instruction for the competing architecture offices. By means of this chain of displacements the primum movens articulated and narrowed down or delimited the needs of an entity, whose identity proved to be rather unstable later on in the process and had until then never been more than a projection.

As regards the set of future inhabitants of the Kalkbreite estate, thirty to forty people intending to move in met for a series of workshops to discuss the size of rooms and flats, how their internal organization might look as well as how the shared and public spaces might be organized. This workshop series, which was recorded and evaluated by the housing cooperative's building committee, can be regarded as a first translation. In a second translation, the findings that had been elaborated on the basis of the workshop series were transformed into no less than nine pages of spatial specifications. This exhaustive listing, which constituted one of the brief's most essential parts, contained binding instructions (as regards, for instance, the size of the intended flats), indications (as regards, e.g., noise protection) as well as inspirations (as regards, e.g., the concept of room sharing). Hence, by means of two displacements – the modification of the initial entity's form and state from a group of 200 future residents into a list of spatial specifications – the primum movens achieved a position as spokesperson for the entire group of future residents.

The rise of a controversy

As described in the section outlining the history of the Kalkbreite, only shortly after the competition, a cinema operator brought into question 'some of the gains of the previous stages' (Callon, 1986, p. 224) – for example the smooth façade and the clear structure of the floor plans of the flats located above or next to the cinema. To use Callon's terms once again, a 'dissenting' future business tenant 'denied' the way it was represented during the competition. By disapproving of the commercial units' floor plans and sections as displayed on the schemes of the winning design, it refused to accept the way it was represented within the competition's outcome and (consequentially) denied the way it was represented throughout the competition, that is, it contested the detailed list of spatial specifications featured in the brief.

At this point we would like to take the focus away from the Kalkbreite case and try to provide a synoptic view of the whole set of cases that we studied (Table 9.2). As a matter of fact, all cases studied can be addressed as either a subset or a variation of the Kalkbreite case: In all thirteen cases a set of potential future inhabitants (and, if applicable, a set of future business tenants) has been translated into a detailed list of spatial specifications. Furthermore, in all the cases we studied, the primum movens succeeded in achieving a (temporary) consensus, which became manifest in the winning design. In what follows, we will be attentive to the processes of translation contesting that consensus and the alliances it implies. In fact, in five of the thirteen cases we studied, the equilibrium that had been established by means of the competition was severely destabilized only shortly after. But while in the Kalkbreite case the dissent came in the form of an 'offer you cannot refuse', the dissidents in the other four cases were far from being embraced.

In the case of the conversion of a former mill in St.Gallen (see entry eleven in Table 9.2), a listed building had been transformed successively into a set of plans displaying which of its components were to be preserved and which could be modified or removed. Hence, by means of a series of expert studies a highly heterogeneous entity comprising construction materials from various centuries (the mill had been built at the beginning of the seventeenth century and reorganized several times throughout the following centuries) had been translated into a tame ally, easily controllable by the primum movens. Obviously, the outcome of the expertise directly found its way into the competition brief, subsequently served as an instruction for the competition architecture offices and consequentially had a considerable impact on the winning design. Yet, shortly after the jury's final decision, an important building component, which had until then been classified as 'not to be preserved' pushed for renegotiating its identity: when examining it, craftspeople excavated precious seventeenth century construction material, which had been hidden under various layers of 'unworthy' subsequent amendments. This renegotiation obviously destabilized the consensus that had been established throughout and by means of the competition. In practical terms it meant that the organization of many internal areas had to be significantly reworked.

In the case of a new construction of a housing complex in Zurich (entry thirteen in Table 9.2), the plot's soil disturbed the competition equilibrium when further test drillings for foundation work unearthed much more subterranean water than the (temporarily) fixed soil identity (which had been shaped by means of geological expertise) had provided for. This reshaping of identity resulted in a major reformulation of the winning design. The shift in the identity of the plot's soil meant that costly foundations became necessary, which in turn meant that the winning design clearly exceeded the defined construction costs. The balance between economic requirements and architectural considerations that had been reached throughout the competition had been destabilized as additional expenditures for foundations were compensated with savings as regards the façade and the choice of materials for the interior.

In a further case that concerns a new construction of a housing complex in Zurich (entry twelve in Table 9.2), an objection by an immediate neighbour shortly after the winning design had been published challenged the temporarily fixed identity of the latter. An enforced reduction in the floor area of the attic led to enlarging the lower floors in order to retain the intended gross floor area. This, however, affected the distance between props, which not only posed a problem in regard to the planned underground parking (the new distances interfered with the intended size and distribution of the parking lots), but also had a significant effect on the floor plans of the flats. The consensus that had been painstakingly established during and by means of the competition as – again – users' needs and wishes had been translated into detailed lists of spatial specifications broke up as these lists and large parts of the design that had been developed on exactly this basis became obsolete.

Finally, in yet another case that concerns a new construction of a small mixed-use complex, in Küssnacht am Rigi (entry two in Table 9.2), the identity of the planning and zoning code, which was negotiated in council exactly at the time when the competition was prepared and executed, did not change in the way the competition organizers expected. This necessarily opened a controversy right after the competition since the winning design had been developed on a then invalid basis, violating the planning and zoning regulations as they had come into effect in the meantime. Hence, the dissenting planning and zoning regulations, obviously, threatened and destroyed a large part of the gains of the previous stages as the winning entry literally fell apart, putting the whole building process on hold.

Conclusion

As we have seen, architectural competitions and their outcomes (that is, the winning entries) have to be considered as (temporary) equilibriums based on the translation of various, often highly heterogeneous entities. We have furthermore seen that these states of equilibrium can be of a rather fragile, vulnerable nature since the negotiation of entities' identities is a continual process. The closure of controversy achieved throughout the competition is not necessarily once-and-for-all; the involved entities may still break open the negotiated delimitations of their identity after the competition.

Against this background we argue that many clients organizing architectural competitions are caught in a dilemma. On the one hand, they try to delimit the identity of the entities involved as far as possible in order to tame as many potential imponderables as they can and to increase planning security. On the other hand, renegotiations of the identity of (at least) one of the entities involved may become necessary, required or desirable after the competition due to shifting circumstances. These shifts in circumstances may produce opportunities for improvement (as in the Kalkbreite case) – they may, however, also produce serious threats to the consensus achieved (as in the other four cases mentioned above). When trying to or having to take such changes in circumstances into account, a (too) narrow or (too) robust delimitation of the respective entities' identities may pose a significant problem. The strict determination of entities' identities as established during and by means of the competition may prevent the client from seizing the unexpected chance that would arise from a shift in an entity's identity afterwards or – in a more general sense – from adequately responding to emerging challenges or threats.

To conclude this chapter, we would like to put forward the hypothesis that executing an architectural competition and integrating it into the whole building process is essentially about finding a balance between determining the involved entities' identities as precisely as possible and at the same time providing an adequate scope for their renegotiation. Referring to De Certeau (1988) we would like to ask whether the strategy to narrow down the scope for negotiations regarding the identities of all entities involved as far as possible can be considered an adequate way of dealing with the complex nature that many construction projects exhibit. Or whether the latter call for more 'fluctuating' approaches or tactics – in particular during the preparation of an architectural competition – which dare to leave scope for the renegotiation of the identities of certain entities involved after the competition. In line with Van Wezemael and Silberberger (2016), we would argue that Callon's (1986) moments of translation are perfectly suited for elaborating the inadequacies of a conduct that aims at locking entities too tightly into place thereby ignoring the fact that the world we are living in (and planning for) time and again produces unexpected, unforeseen events.

References

Callon, M. (1986). Some Elements of a Sociology of Translation: Domestication of the Scallops and the Fishermen of St. Brieuc Bay. In J. Law (Ed.) *Power, Action and Belief: A New Sociology of Knowledge*. London: Routledge and Kegan Paul, pp. 196–233.
De Certeau, M. (1988). *Kunst des Handelns*. Berlin: Merve.
Koolhaas, R. and Mau, B. (1995). *S, M, L, XL*. New York: Monacelli Press.
Melhuish, C. (2013). *Luis Vidal + Architects – From Process to Results*. London: Laurence King.
Menz, S. (ed.) (2009). *Drei Bücher über den Bauprozess*. Zurich: vdf Hochschulverlag.
Muhm, A. (2014). *Ein multifunktionales Modell des Projektmanagements im Hochbau*. Wiesbaden: Springer Verlag.
SIA (2001). *Regulation SIA 112*. Zurich: Schweizerischer Ingenieur- und Architektenverein / Swiss Society of Engineers and Architects (SIA).
Van Wezemael, J. and Silberberger, J. (2016). Emergent Places: Innovative Practices in Zurich, Switzerland. In Y. Rydin and L. Tate (Eds) *Actor Networks of Planning: Exploring the Influence of Actor-Network Theory*. Oxford: Routledge, pp. 175–185.

10 Architecture as process
OJEU tender and procurement without design

Torsten Schmiedeknecht

Competitions are regarded within the architectural world almost as motherhood and apple pie issues, concepts that nobody could reasonably question, presented as good deeds in an unkind world. They are understood as an expression of disinterested commitment to quality. It's hard to argue in the face of the many major examples of competition winning buildings that have served to define the architectural history of the 20th century. But it is worth interrogating at least for a while, the received wisdom that competitions are uncomplicatedly good things.

(Sudjic, 2006)

Introduction

Open and anonymous architectural design competitions are often seen as a hot bed of architectural innovation, a field in which new ideas are tested and investigated and within which architecture can be explored for the sake of architecture (Lipstadt, 1989).

In contrast to continental Europe, where design competitions as a procurement method for architecture projects are routinely employed, Official Journal of the European Union (OJEU) procedures are a commonly used procurement tool in the United Kingdom. The procedure is usually carried out in two stages. After a project is advertised, architects, often in teams with consultants, fill in and submit a Pre-Qualification Questionnaire (PQQ). Information usually requested includes, for instance, statements on practice history and size, on quality assurance and financial standing, and a design statement relating to the advertised project. The client subsequently draws a shortlist, and candidates are interviewed in order to find the most suitable team for the project.

In the competitive interview of an OJEU process without design component, a practice's track record, but also, and perhaps more importantly for prospective clients, the evidence of how this work has been received by clients, the press and perhaps the broader public, is the subject of the assessment of design quality, and in that sense represents the replacement of the specific design submission to a competition.[1]

In previous research I have investigated the role anonymous design competitions play in instigating and maintaining architectural discourse (Schmiedeknecht, 2013; 2011; 2007; 2005).[2]

The value of design competitions for the culture of architecture via the shared platform of competition entries, the options presented to clients by the different submissions to competitions, the value for the development of practices regularly participating in competitions, and the fostering of young practices, is undisputed. However, (open

152 Torsten Schmiedeknecht

and anonymous) design competitions are not always entirely unproblematic. The lack of collaboration between client and architect in the brief development stage, for example, or the fact that clients can be left, after the jury has gone, employing an architect they cannot establish a good working relationship with, can lead to potentially problematic project developments. This chapter explores aspects (such as the possibilities for client / architect collaboration; and the question of specialization vs. variety) of the procurement of architectural services for publicly funded projects via a two stage OJEU process *without* design component.[3] This is compared to procurement by open and anonymous design competition. The main question the chapter is asking is whether or not OJEU procedures *without* design component can be beneficial for clients, for architects and for architecture, and if and how they can provide a positive alternative to procurement via design competition. In order to address this question the chapter is structured into six sections: Competitive procurement methods and architectural discourse; Design competition or OJEU – variety or specialization?; Finding the design team; OJEU interviews; Collaboration between client and design; and the conclusion, Architecture as process or architecture as product. Two case studies of buildings in Liverpool, UK, recently completed on Hope Street, and procured via OJEU, served as the research vehicle: the new Everyman Theatre by Haworth Tompkins Architects, completed in 2013 and winner of the 2014 Stirling Prize, and the 2014 refurbishment as well as the new extension in 2015 of the Philharmonic Hall, by Caruso St John Architects.[4]

A series of interviews over the course of 2015 were conducted with representatives of the clients for both projects, with the architects, the contractor and the planners from Liverpool City Council. In addition a further interview was held with the architect Cindy Walters, member of the Stirling Prize jury in 2014.[5] The interviews contained two main sets of questions. The first set related to the individuals' general experiences with procurement, in particular regarding their opinions and perceived differences between OJEU and design competitions. The second set then related to the interviewees' respective experiences with the actual projects for the Everyman Theatre and the Liverpool Philharmonic Hall.

Competitive procurement methods and architectural discourse

> Competitions have survived these many centuries because they reaffirm architects' social identity as cultural producers, 'analogous to sculptors, not to stone masons' (Ware, 1899), in short, as relatively autonomous artists.
>
> (Lipstadt, 2006)

> Obviously you need a culture of architecture to support a successful design competition. OJEU, the idea of tendering for a job ostensibly removes this culture. If it is led by a project manager, or the procurement department of a city, a cultural imperative has been effectively removed from the process.[6]

As Lipstadt states, design competitions can have the effect of promoting architects as producers of autonomous artefacts, likening the architect's autonomy to that of an artist. When architects enter a competition, what really is an *application* to win a commission, is also an independent contribution to a discourse on architecture.[7] It expands the architect's role and understanding of their work beyond that of the provider of a service for a client.[8] Winning schemes will be published and discussed in different media and even unsuccessful competition entries will find their way onto architect's web sites, the web sites of institutions running competitions, into journals and magazines, and so on.

Cindy Walters, referring to both her own practice and to her judging the Stirling Prize, sees competitions, whether they are design competitions or two-stage OJEU procedures with design component, 'as currency', serving a number of purposes for the practices involved.[9] If a practice, for example, is trying to enter a new sector or is interested in developing knowledge and expertise on building types they have not been involved with in the past, a submission to a competition serves as a useful research vehicle, both for the practice as a whole, but particularly for those members of staff who are directly involved. Designing for a project in a new sector can therefore, according to Walters, be 'very refreshing'. In this respect, any competitive tender process which includes a design component, is also a contribution to a discourse on architecture. Walters also states that most projects awarded the Stirling Prize will have started their lives as part of some kind of competitive process:

> The Stirling Prize is looking for the project that has done most to further the discourse on architecture in the particular year. Excellence is one of the criteria, as are many other things, but actually on balance it is the project that has done most to further that discussion. Most projects that win the Stirling Prize have started as competitions, almost all of them, so competitions and awards are linked.

How does this compare to an OJEU tender process where there is no design proposal involved at any stage of the bidding – as was the case for both the Liverpool Philharmonic Hall and the Everyman Theatre? What kind of cultural context can OJEU procedures be placed in and do they further the discourse on architecture at procurement stage – which is arguably what design competitions do – for the practices involved, the broader architectural world, and for the clients and subsequently the general public?[10]

While the process of finding a winner for an architectural competition is undoubtedly one of intense communication and exchange of opinion between (mainly expert) jury members, it represents an *internal* discourse, taking place and steered mainly *within* the architectural profession.

As Larson (1994) states:

> The traditional identity and the traditional legitimations of the architectural profession also appeal to the ideology of art. [...] Here, I consider as 'discourse' all the statements that a particular category of agents issue in a specific capacity and in a definable thematic area. The entries in a design competition [...] are such statements. In architecture as in all professions, discourse is open only to those who know 'how to speak'.

Depending on the composition of the interview panel, a broader discourse between clients and architects, in which both parties 'know how to speak' (also on architectural issues), is, on the basis of a practice's previous work and of the project at hand, potentially being enabled at the interview stage of an OJEU process.

154 *Torsten Schmiedeknecht*

Unless the proposal to undertake a new building project is highly controversial and sparks public resistance or debate, discourse involving the general public is usually absent prior to an architectural commission being made. In contrast, the results of and schemes submitted to design competitions are, particularly in Northern (continental) Europe, often subject to publication and exhibitions. These exhibitions are open to both the general public and participating architects, and are potentially the venue for an overlapping discourse between the profession and the public. It is very rare, however, that the public, even for projects financed through a proportionally high amount of public funding, are aware of the decision making / commissioning process within an OJEU procedure that does not include a design component. Although the Everyman Theatre and the Liverpool Philharmonic informed the public via press releases and their web sites about decisions and progress made, the public were offered the information as a fait accompli. In an exhibition of competition results, although the public may not be able to influence the winning scheme, they are invited to compare options, and thus enter the architectural discourse of their culture.[11]

When entering an open design competition for a specific scheme practices will normally submit a new, previously unseen design. In an OJEU process *without* design, practices need to respond to specific questions, and speculatively state how they would deal with certain aspects and difficulties to be expected in a project of which they don't really know the programme. Statements from the PQQ or the final tender document are then often subsequently picked up in the interview in stage two. An important part of the submission will therefore be of a retrospective nature, that is, highlighting a practice's past achievements, an aspect that is entirely absent from open design competitions.

While a design competition gives a client, or their representative(s) the opportunity to compare different proposed solutions, inexperienced or first time clients often do not have the expertise to adequately assess and judge architectural submissions and therefore rely on the help of the expert members of a jury.[12]

As stated above, in an OJEU process track record and past client reception replace the design submissions asked for in a design competition. Haworth Tompkins' tender submission for the Everyman Theatre, for example, included, in addition to their track record of successfully carried out projects, testimonials from previous clients and quotes from the architectural and broadsheet press, with particular reference to the theatre sector, commenting positively on their work.

The contribution to a discourse on architecture in a design competition is based on the design work submitted to the competition, and the discussion takes place first *internally* in the jury, and then in a more public way via exhibition and publication. The discourse is started prior to a building being built; it is based at that point on paper architecture. This kind of discourse cannot take place in an OJEU procedure with no design.

Design competition or OJEU – variety or specialization?

> At the same time they [competitions] serve as an educative spectacle, an exchange of talents, ideas, and forms, as a traveling exhibition of skills. [...] the open competition brings out the opportunity to make architecture for its own sake.
>
> (Lipstadt, 1989)

Procurement, because it is a really big part of the process is very important, and it's the way that all our buildings start their lives. So much of the profession is involved in the frantic scramble for visual wins, and the substance of what we're doing is in danger of getting eroded.[13]

One of the perceived advantages of open anonymous design competitions over other procurement methods is that of the nurturing and fostering of young practices. Young architects competing against established firms not only stimulates discourse but also has an educational purpose for architects tackling a particular task or building type for the first time.

Do young and relatively inexperienced practices have a chance to compete in the same way for prestigious or large commissions via OJEU, or does the pre-qualification process effectively shut the door on anyone lacking experience with the design of the specific building type of any particular OJEU tender? And what, in that case, could be the possible consequences for architecture?

Open competitions potentially open the field and give architects the opportunity to test themselves in untried sectors, gaining experience in the design of different building types, and even an unsuccessful competition entry might subsequently find its way into a practice's application for an OJEU tender. Does an OJEU tender procedure with PQQ have the opposite effect and favour only practices with a significant track record in the relevant sector? And would any of the above have a positive or a negative effect on architecture, that is, is diversity always a good thing, and does, conversely, specialization lead to limitation? Or could the opposite be the case with regards to OJEU procedures, namely that experience leads to better results?

Considering the work that Haworth Tompkins and Caruso St John had carried out at the time of tendering for the Everyman Theatre and the Liverpool Philharmonic Hall, respectively, it is evident that both practices brought experience from the relevant sector.[14] In Steve Tompkins' words, although his practice's theatre experience was relevant:

> it was an intensification of the conversations and methodologies that we had been developing over the last twenty years. Most people think that we are 'new theatre' people, which in fact we're not. So we had to apply methodologies of working with existing situations and existing contexts to a new build. It felt like quite a different process.

However, the Everyman, resulting in the winning of the Stirling Prize, proved to be the culmination of years of experience working with and fine-tuning a particular type of programme. In the context of the Liverpool Philharmonic Hall, Adam Caruso points out that his practice is not famous for their technical wizardry, but that:

> all the visual arts and museum projects we've done, they were all technically very demanding, whether they're existing or new. And I think we have a huge amount of experience in installing or changing services so that they're contemporary in a way which doesn't have a big impact on the fabric of the building, whether it's a new building or an old one.

While neither practice is 'new' or 'inexperienced' any more, it is worth remembering that Caruso St John's breakthrough came when they won the two-stage design competition for the Walsall Art Gallery in 1995. The practice has and is continuously participating in a variety of competitive processes to secure work, which includes design competitions, both in the UK and in continental Europe. Caruso suggests that his firm would most probably not have been as successful, and in particular would not have been able to secure work in Sweden, Germany and Switzerland, had they not participated in design competitions.[15] Steve Tompkins, on the other hand, has a very different view of the value of design competitions, and maintains that the Everyman could not have resulted out of such a process:

> it [the Everyman] validates the fact that you can't go in to a project like that with a preemptive strike and go 'we're really hot' and this is it and we don't care what you think because you employ us, you pay us to have fantastic ideas and for this building to look brilliant and to win those awards for you.

In addition to Caruso St John, Liverpool Philharmonic's shortlist included Allies and Morrison, John McAslan & Partners Ltd, Bennetts Associates and Witherford Watson Mann Architects. Of the five practices shortlisted, Allies & Morrison and Bennetts Associates brought the weightiest experience in concert venues, followed by John McAslan & Partners and Caruso St John.[16] Witherford Watson Mann, while having individual experience from work carried out in previous practices, hadn't delivered a comparative scheme at the time of tender. However, all of the practices on the shortlist had significant experience of dealing with existing, and more importantly, listed and heritage environments.

For the Everyman the PQQ asked for 'details of relevant experience of completed (in the last 3 years), or in progress, publicly funded arts projects demonstrating your understanding and commitment to performing arts and leisure, particularly theatre'. Furthermore there was a separate score for heritage project experience. Including Haworth Tompkins, the shortlist was Arts Team, Ian Ritchie Architects, Keith Williams Architects and LDN Architects, and all of the shortlisted practices brought significant experience in the theatre sector.[17]

Arguably, and with the exception of Witherford Watson Mann, or to some degree even the winning practice Caruso St John, the selection criteria for both shortlists made it virtually impossible for any practice without relevant experience of the proposed building type to compete. The fact that Caruso St John and Witherford Watson Mann were shortlisted in the first place is due to the professional insight of Emma King, Liverpool Philharmonic's Capital Development Director:

> When procuring an architect for a project, it is not necessary that they have done the same sort of work before, it is not important that they have a pedigree. But they should have an agenda, a motivation and a way of working that will provoke an interesting response.

Open design competitions could have enabled the clients to choose from a more varied response on paper. However, and while the procurement route chosen in both cases prevented 'the opportunity for the discovery of new talent' (Lipstadt, 2006), both clients had the benefit of developing their projects in dialogue and collaboration with architects possessing the relevant skills and experience to deliver culturally sensitive and technically complex building projects.

Finding the design team

> Philip Johnson and Jean Prouve were able to persuade their fellow judges in Paris to go for one of the most radical of several hundred submissions for the building of a cultural complex next to the old Les Halles. When they opened that envelope, they found that they had entrusted the most high-profile, and one of the most technically demanding new buildings in France since the Eiffel Tower, to an entirely unknown Italian called Renzo Piano, and Richard Rogers, an Englishman who had previously built nothing larger than a seaside house for his father-in-law.
>
> (Sudjic, 2006)

> Where the architecture should come from? I have a belief, that getting the right architect, to ask the right questions, can surprise everybody. So it's not about the social or economic context framing the piece of architecture, it's the architect and therefore the product, framing a different set of questions. I think that's an important principle for how I have gone about procuring and working with architects in the past, that I have a belief in their skill set, and that I have a belief in the work that they do, which will surprise everybody, and hopefully delight everybody.[18]

When the architectural services for a project are procured via an anonymous design competition, the client, represented by a jury, is first and foremost making the award of the project based on the quality of a design proposal. The anonymity of the competing architects should normally suggest that the client and the jury are unaware of the winning scheme's authorship. It is therefore clear that, at competition stage, an architectural solution to a given problem is sought, rather than, as is the case in a two-stage OJEU procedure *without* design, a team that will subsequently develop one or several solutions to the client's problem, that is, the brief. A submission to a design competition also does not necessarily always include the contribution of, for example, structural engineers or mechanical & electrical consultants. During an OJEU process, however, they are often appointed at the same stage as the architects or 'lead consultants', that is by submitting a tender document followed by an interview with the client and their representatives. In the case of both the Everyman and the Liverpool Philharmonic Hall, and as is common practice, tender submissions were invited from architect lead design teams, including all the relevant consultants.[19]

One commonality between design competitions and OJEU tenders can be that of a client actively inviting (competitions) or encouraging (OJEU) practices to take part in the competitive process.[20]

At the beginning of the Liverpool Philharmonic project, for example, architects were alerted by the organization that an OJEU tender would imminently be published, and that the Philharmonic would be interested in their submission. As part of these invitations or encouragements, in invited competitions there might be a remuneration for participating practices to reward them for their efforts, and in general there is prize money to be distributed. OJEU tenders are normally entered by practices for free, but the process is perceived to be less labour-intensive than the development of a fully fledged design proposal.

Emma King states that her experience of the architectural press was helpful in 'talking about practices', in order to add to the 'strata of mainstream practices', because 'it's always interesting to try and promote people who are emerging, or about to emerge, to take part'. For King, part of trying to diversify the eventual shortlist was to make the architectural press aware that the project was going live, in order to get coverage in trade journals and to raise architects' awareness of it.

However, recalling the initial process of generating interest in the project, she remarks that she was 'surprised, moving from an organization like Arts Council England, which due to its role in being able to provide significant support through lottery funding, saw architects keen to be attached to the name, how hard it was attract a really good field'.

What are the advantages for a client, particularly if they are fully or partly publicly funded, to pursue the procurement of a project via invited tender and competitive interview rather than via design competition? King is very clear as to where her preferences regarding procurement procedures lie. She recalls an example of a competition where she was involved on behalf of the Arts Council as funder, and where the winning project was subsequently abandoned because it suited neither the client's requirements nor the site. The winning project, in her view, was 'an object, it was part of that period of time when the iconographic cultural centre was the Zeitgeist of the day'. King argues that this early experience had a big influence on forming her opinion on the procurement methods of publicly funded architecture. In her view, design competitions are unsuitable as instruments for the procurement of design services, in particular when part of the project is the refurbishment of an existing building, as a concise brief for a design competition dealing with technical improvements for a concert hall, as well as with sensitive listed and heritage building issues, is difficult to achieve at procurement stage.

Another potentially problematic issue with design competitions is the distant relationship between jury, client and architect. Evidently, design competition juries can be dominated by expert members, that is architects, as they are in possession of the field-specific knowledge and experience that a first time client cannot possibly have acquired.[21] And because there is no dialogue between architect and client / jury prior to the award of the winning scheme, clients are potentially faced with a design team of which they know little or nothing. Considering that any development of a project, post competition, in all probability takes place in the absence of the expert jury members, this can lead to complications and can potentially put a strain on the client–architect relationship at the early stages of a project.

In the case of the Liverpool Philharmonic, the panel for the competitive interview sessions was composed of longstanding representatives from the Board and senior management from within the organization, and the relatively newly appointed King, who had previous experience of the OJEU process from her time at the Arts Council, where she had been responsible for the London estate and had commissioned projects through the estates programme. Not only did King, a qualified architect, have expert knowledge of the process undertaken, she had also effectively written the OJEU documentation and was leading the project all the way through to completion as the client's representative.[22] All of the panel members represented an interest in the project beyond the outcome of the interview. This is in contrast to Sudjic's observation that (design) competitions 'are understood as an expression of disinterested commitment to quality' (Sudjic, 2006).

OJEU interviews

Steve Tompkins is adamant that the interview is a vital and not to be underestimated part of the appointment process – however small in percentage terms its contribution might be to the overall scoring of the applicants – as it represents the moment in which the architect (or the whole design team, as was the case with the Liverpool Philharmonic Hall) and the client, can have a dialogue around set parameters. The panel seeks to assess not only the expert abilities of a team, but also, and perhaps more importantly, whether or not the 'chemistry' between applicant and client looks to be promising with regard to a fruitful collaborative process. At the Everyman, Tompkins and his team were asked to give a presentation of the practice's work and approach, but also to each name and describe one of their favourite buildings:

> It was Gemma, Deborah and Robert wanting to understand the individuals they would be working with, trying to understand what motivated them and what their DNA was going to be like and what the chemistry of the team was going to be like. Could they work with these people? You find this much more in our experience with theatre clients, because they understand that the effectiveness of what they do is about working relationships and it's not about a preconception.

In the Liverpool Philharmonic interview, the whole design team, that is architects and consultants, were placed on the existing stage, facing the auditorium, both in order to test the team's performance in an unusual setting, but also to give those members of the organization who were not part of the panel but interested to meet the applicants, the opportunity to observe and listen to the procedure.[23]

The probable future chemistry between client and design team was also being assessed in the interviews for both projects via a clarification over who on the architect's side was going to be running the job, and how present they would be in the process, or how much access the client was going to have to the practice directors over the duration of the project. Furthermore, the Liverpool Philharmonic interview panel enquired about how the architects would assist in raising funds, how they anticipated helping to overcome anxieties on behalf of the stakeholders who still remembered (less fondly) the last period of transition and refurbishment of the hall in the 1990s, and how the architects could help to build confidence among the stakeholders. With reference to the tender documents, architects were also asked about the split in the project between working within heritage environments (the foyer and hall) and the desired (on behalf of the client) strong design response, that the 'totality of the site calls for', and 'what [...] strengths [they] would play to / what needs to be addressed'. One important factor was the technical nature of the project, that is, 'what will a successful project look like from a design and technical perspective'? Adam Caruso explains that there had been:

> a very, very detailed list of, you could say, programmatic things why they wanted to do the project. The fact that there were no mechanical services functioning in the building, let alone in the auditorium. There were also some acoustic issues, and then there was the whole stage machinery – very practical things that were problematic and had to be dealt with.

The Philharmonic were also concerned that prior to the current project a lot of problems in the daily operation of the building had been caused by assumptions about audience behaviours based on patterns from the 1930s and the 1990s, and the architects were asked 'how can you help us to ensure that we get it right this time and avoid having to repeat this exercise in fifteen years time?' One of the key ambitions of the project was to make the hall more commercially viable while 'respecting and preserving the special nature of the hall and stage'. In that context the architects were asked how they could help the organization 'to prioritize our investment and make difficult choices'.

Robert Longthorne has a different view as to the role the interview played in the procurement process. According to him, 'the interview process was only a small part of the actual overall scoring. [The interviews] gave us an opportunity to meet people and to address any queries there were from the tender process.'[24]

The aspect of cost and fees has to be considered in both the scenarios of open competitions and OJEU process. While the fees might not be the overall deciding factor in an OJEU process (although the RIBA survey suggests that they often are), particularly with clients procuring buildings for culture, they are however decided at the stage of 'winning' the bid. In an open design competition the prize money conventionally represents a percentage of the overall project fee to be expected, and is, at that stage, not negotiable.[25]

It is evident from the above, that the questions asked in the interviews would have been impossible to address in the context of an open design competition. Considering the complexity of the issues to be addressed in both projects, combined with the need to assess the client / architect chemistry that both King and Tompkins considered to be all important on their list of priorities, to procure either project via an open and anonymous design competition would have been a very difficult undertaking indeed.

Collaboration between client and design team

> In the end, however, the competition's greatest benefit for the architect was the opportunity for creation unfettered by client control.
>
> (Lipstadt, 2006)

But Deborah and Gemma, you know, they embraced the Everyman project like nothing I have ever seen before. They really gave everything to it. And the hours Emma King is putting in to make sure that she is representing the Philharmonic, it's mind blowing, the effort she is putting in, absolutely incredible.[26]

One of the things we did at the beginning of the consultation process was to ask for the consultation meetings to be documented and we would then comment [...] and the final version would go into the records. That was always very good. Because nuances can easily get missed, so it's important to make sure that they are picked up and carried forward. I think that process actually did work extremely well.[27]

Lipstadt suggests that architects who enter competitive procurement situations prefer to be creative without client interference, and that design competitions provide this platform. Two-stage OJEU processes with no concise brief or schedule of accommodation in stage two, and in particular those where the delivery of a 'design' is neither encouraged nor discouraged, are in that respect less desirable to architects. They can lead to architects feeling obliged to deliver some form of speculative proposal based on a yet-to-be-defined brief, because, as Steve Tompkins explains, clients may just say 'we're not looking for a design, we're looking for an approach, so could you supply five A1 boards illustrating your approach?' If a practice is subsequently given four weeks to illustrate their 'approach', it seems that this is another way of saying that they are participating in a design competition, and 'they [the clients] give you just long enough as well to actually think everyone else must be doing something'. This way of operating can then be perceived by architects, because of the potential amount of unpaid work that is required to 'comply', and the lack of communication between client and architect regarding what is actually asked for, as slightly 'abusive' (Tompkins), and therefore perhaps not the best way of starting a collaborative working relationship.

In contrast to what Lipstadt states is the creative benefit for architects in doing competitions, Barrett and Stanley (1999) have argued that 'without engaging end users in the process, an important stimulus to the creative process of design is lost'.[28] In Haworth Tompkins' case, the architects say that their whole working methodology is based on careful communication, that is listening to and understanding the client's and end users' problem before attempting to develop a proposal. At brief development stage the architects were an integral part of formulating the Everyman programme and of the decision to develop the site on Hope Street, including the purchase of a neighbouring building, to be able accommodate the offices for both the Everyman and the Playhouse on one site. At the Liverpool Philharmonic Hall, the proposal to demolish the existing back of house and to replace it with a new building on the old footprint and using part of the existing steel frame was equally a result of the communicative process between architect and client. For the Everyman, numerous public consultation events were held over the duration of the project, in order to ensure that the qualities of a treasured and loved institution would not be lost in a new building, and therefore input of the end users was actively sought at various stages during the process.

In the cases of the Everyman Theatre and the Liverpool Philharmonic Hall, two examples, both sensitive and crucial for the success and acceptance of the schemes, perhaps stand out with regards to the close collaboration between architect and client / stakeholders: the design and resolution of the Hope Street elevation for the Everyman, and the colour and lighting scheme of the Entrance Foyer and the Grand Foyer Bar for the Liverpool Philharmonic Hall.[29]

According to Robert Longthorne, the idea to have 105 life-sized water-cut portraits of contemporary Liverpool residents on moveable metal solar shading panels on the front façade first originated from Steve Tompkins, 'but it came out of a number of things and influences we'd been looking at'. One of the questions everyone was concerned with was how to express the company's ethos, and how to express *Everyman*. Tompkins states that the façade had to:

> have that civic presence, and on the other hand it had to be part of the street wall, so it was treading a fairly interesting tightrope. It had to perform a very specific technical task as well, which is sun shading because it is a West facing elevation.

Figure 10.1 The new Everyman Theatre in Liverpool: principal façade towards Hope Street.
Source: Philip Vile.

Figure 10.2 Liverpool Philharmonic Hall: the Entrance Foyer.
Source: Author.

Figure 10.3 Liverpool Philharmonic Hall: the Grand Foyer Bar.
Source: Author.

> All of these things started to synthesize into a solution but the idea of repeating ranks of human scale images was just a gut intuition from very early on in the scheme. It came from looking at medieval cathedral façades, from Lowry's paintings of ordinary people, some quiet manifestations of humanity.
>
> It is a representation of 'Everyman', the anonymous inhabitant of the community. I think we erred towards a very figurative representation of those individuals.

Longthorne remembers the analogies to cathedral façades and the comparisons made in discussions:

> Steve came up with a number of references: the West front of Wells Cathedral, and something that Gemma picked up on [...] were the fronts of cathedrals in France, where [the] ordinary people of the area [...] play a more important part. In Wells you got the great and the good at the top and the ordinary people underneath, [...] in France the balance is different and you very much have the ordinary people. So that [...] idea represented the people of Liverpool.

Once the photographer, Dan Kenyon, had been commissioned through a competitive process, photography pop-up studios were set up in various locations, sessions at the Playhouse held, and local communities approached. '[People] selected themselves to be photographed and we photographed anyone who turned up, resulting in a few thousand photographs because there were several of shots of each person' (Longthorne). Steve Tompkins, Gemma Bodinetz and Dan Kenyon then selected the best shots of each person and started to look at the composition of the 105 images needed for the façade.[30]

A challenge to the architect–client relationship for the Liverpool Philharmonic Hall was the colour and lighting scheme for the front of house. Equally developed in a collaborative process, it illustrates Emma King's preference of OJEU *without* design over design competition as a procurement method, and Adam Caruso's suspicion that 'it's really crazy doing a design competition for something where two-thirds of the budget is in the existing building. And how do you compete on a project which is 30 per cent services?' Both King and Caruso remember 'quite difficult conversations within the organization', and the colour scheme was not entirely undisputed. When the Liverpool Philharmonic Hall was refurbished in the mid-90s, most of the public interiors had been turned magnolia and Caruso says the architects instantly developed visions of what they thought the interiors must have originally been like. In order to achieve more clarity on this, paint scratch tests were carried out, but, according to King, they had not been entirely conclusive. Not being able to find conclusive evidence of the original 1930s colour scheme, Caruso St John turned to other Rowse buildings in Liverpool, like the Martins Bank, in which they saw a 'robust, vibrant and quite radical' colour scheme. Caruso recalled saying to Emma King 'I'll show you the colour scheme in a couple of weeks, but be warned'. King's recollection is that, because the original colour scheme was not available, and the architects hadn't necessarily experienced that before when they had worked in existing buildings, they also turned to other interiors typical of the time, and not necessarily linked to Liverpool, such as the interiors of a house by Adolf Loos, the Skandia theatre by Asplund, and a more anonymous image from a 1930s cinema in Newcastle as a reference for the box office ceiling. This part of the project thus was about creating an atmosphere that would combine what the original building would have been like with an appropriate response for a festive contemporary concert hall, and the transition from one of Liverpool's more important streets via a grand foyer into the actual hall. King remembers the process as collaborative, but driven by the architects, which is, in her view also why they were appointed:

> I think we have been really brave actually. We put our trust in them and more or less went with everything they proposed, although we were questioning and challenging them all the way. Inevitably in a project like this, in a much-loved listed building, there was some heavy water to get through, not least in terms of the colour scheme. But I think it's really clever and quite brilliant, borne out of the fact that the paint scratch tests which took us right back to 1939 revealed that the original interior was probably pretty unpopular and didn't last long.

The hope for the Liverpool Philharmonic is now that, and as they had articulated in the OJEU interview with the architects, they did 'get it right this time and won't have to repeat this exercise in fifteen years time'.

Architecture as process or architecture as product

While it is not is not the purpose of this chapter to express any preference for one or the other procurement method, the findings suggest that projects procured via OJEU *without* design can be process driven and allow for a more collaborative relationship between clients and architects. Particularly in cases where the desired final product is 'open' at the time of appointing the design team (Everyman), or when it is difficult to base a brief on the existing problem (Liverpool Philharmonic Hall), it is important to, as King states, 'bring the right kind of people to the table, which means progressing an agenda rather than following an agenda'. If we consider Larson's (1994) observation that

> Important [design] competitions are important, first of all, by the nature of the project; but they are also important for the *cognoscenti* (which includes the sponsors and their representatives), because their outcomes can change the position of the players in the specialized field of architecture

it is easy to see the temptations and difficulties arising from the potentially elevated position an architect might find themselves in after having won a commission via design competition. In contrast, the emphasis in the procurement of the architectural services in the two projects discussed here, was on finding design teams willing to engage from the very beginning with the particular dynamics and the people of the respective organizations, in order to first develop the project brief and to define the problem, and to subsequently find an appropriate architectural answer. Arguably, and as demonstrated, it is beneficial for a design team to work with an organization if an engagement is enabled from the earliest possible moment. The scenario after a design competition can be much more complicated: for instance the client might not be entirely convinced by the awarded scheme and – after the (perhaps) 'architect-dominated' jury has long gone – the personal chemistry between client and design team may be difficult to establish, particularly because the client potentially doesn't 'know how to speak', that is design team and client are not part of the same discourse.

The story of the Everyman Theatre's façade in particular illustrates a process of collaboration after competition that was, though led by the architects, enhanced through a close working relationship with a client willing to both engage in the proposal but also to be an active part of its development. The process of finding and defining the project solution could not have been part of a conventional design competition. The engagement with the local community via the representation of its people on the façade enhanced a collaborative process, that for Tompkins is 'more deeply rooted and more fundamental' and that he believes is so important for a successful project.

The appointment of Emma King provided the Liverpool Philharmonic with the necessary expertise to guide the organization through the difficult process of finding an architect capable of engaging creatively in the problem at hand without overwhelming a delicate existing heritage environment and its users.

While de facto not suitable to discover or nurture new talent, and perhaps limited in its ability to raise discourse and diversity, architecture as process (OJEU), as opposed to architecture as product (anonymous competition) at procurement stage, can however enable clients and architects to achieve results that will arguably stand the test of time.

It is evident that in the case of both the Liverpool Philharmonic Hall and the Everyman Theatre, the procurement route enabled a process of dialogue and consultation (i.e. the opposite of 'the opportunity for creation unfettered by client control') ultimately benefitting client, architect and end user. In the case of the two projects studied, a positive contribution has not only been made to the two institutions, but also, and perhaps more importantly in the long term, to Hope Street and to Liverpool as a whole. Finally, and on the basis of the research findings, what conclusions can be drawn regarding the suitability of particular procurement methods for any project?

The two projects discussed here were both located in the performing arts sector. The main characteristic that they shared was that both clients had a general idea of what they wanted from their projects at the outset. In neither case was there a finalised brief or schedule of accommodation at the outset, and the projects were thus developed and defined in close collaboration with the architects.

In the case of the Liverpool Philharmonic Hall, the fact that a substantial part of the project was dealing with the refurbishment/enhancement of a listed environment, and that the project involved considerable monitoring and evaluation elements, were also deciding factors in the choice of the procurement route. A conclusion can also be drawn that the positioning of an experienced architect, not solely within the design team, but also as the client representative, enabled a positive outcome for the Liverpool Philharmonic Hall.

There are numerous projects from the cultural sector, particularly in Germany that, in contrast to the schemes discussed here, were procured by design competitions and which received critical acclaim at the time of their completion and, for different reasons in each case, went on to stand the test of time. However, these projects started out with concise briefs and accommodation schedules.

Some, like the Jewish Museum in Berlin (2001) by Daniel Libeskind, opened a new market to a foreign architect, and others, like the Schirn Kunsthalle in Frankfurt (1985), by Bangert Jansen Scholz and Schultes, helped to promote and give young and upcoming architects their first chance at a prestige project. The Münster Library (1993) by Bolles Wilson arguably did both.

To conclude, it is the condition – organisational, physical, political – and also the respective ambition of a project that exists at the outset, rather than the type of building procured, that are the relevant factors when choosing the procurement route for a client and a project.

If the site, brief and general parameters are already formulated for a project and are likely to remain fairly fixed, and a defined schedule of accommodation exists, then open and anonymous design competitions can be a suitable procurement method. Furthermore, if there is an interest, for whatever reason, to broaden the field of competing architects and thus possible solutions, and to give young architects a chance to provide refreshing and possibly radical solutions, then open and anonymous design competitions are equally suitable. As in the example of the Jewish Museum, there is potential additional publicity for the client and project to be gained by a radical design most likely to be generated by architects designing in relative isolation.

If, however, a project brief is not fully formulated, heritage environments or substantial technical demands are involved, and if clients are interested in drawing and benefitting from the architect's skills of involving stakeholders and communities, as much as the architect's past experiences in a particular sector, OJEU without design, as demonstrated in the examples described would be a more appropriate procurement route.

The two procurement methods compared and discussed in this chapter are evidently at opposite ends of the spectrum. While design competitions may still offer the possibility of discovering new talent and thus perhaps 'advancing' architecture, the discourse taking place is, at procurement stage, 'closed'. OJEU procedures *without* design, on the other hand, can offer a more collaborative and client centred approach. Further research in the field of competitions and procurement may yet lead to procurement methods combining the most positive aspects of each of the two routes discussed here.

Acknowledgements

I would like to very much thank the following individuals for their help and the time they made available to me during the writing of this chapter: Adam Caruso, Steve Tompkins, Emma King, Robert Longthorne, Cindy Walters, Raymond Gilroy, Barbara Kirkbride, Chris Ridland.

Notes

1 According to a procurement survey carried out for the Royal Institute of British Architects (RIBA), 'in 2011 the UK architectural profession spent £40m preparing OJEU process bids, equivalent to 29 per cent of the fee earnings they derived from an estimated £138m of OJEU-derived work.' The survey states that 'the types of OJEU procedures most often used are not those […] that architectural practices would prefer to undertake', and that 'the vast majority of skilled practitioners across the country are either excluded from the predominant bid type, the Pre-Qualification Questionnaire (PQQ) process (as a result of their practice size or financial standing), or choose not to participate, citing the high costs involved' (Mirza & Nacey Research, 2012).
2 In contrast, and adding to these previous papers, the opportunities for discourse taking place as part of or resulting from OJEU procedures, for instance, are discussed in comparison to the discourse arising from design competitions.
3 PQQ, followed by full tender with interview.
4 The OJEU notices for the Everyman and the Liverpool Philharmonic Hall were published in 2006 and 2012 respectively. Liverpool Philharmonic Hall was originally designed by Herbert Rowse from 1936 to 1939.
5 Interviewees were: Steve Tompkins and Will Mesher (both Haworth Tompkins); Adam Caruso (Caruso St John); Robert Longthorne (Building Development Director, Everyman and Playhouse Liverpool), Emma King (Capital Development Director Liverpool Philharmonic Hall); Raymond Gilroy (Construction Director, Gilbert Ash); Cindy Walters (Walters & Cohen); Barbara Kirkbride (Principal Planning Officer LCC); Chris Ridland (Principal Planning Officer LCC).
6 Adam Caruso, Director, Caruso St John Architects, interviewed by the author, London, 25 August 2015.
7 See also Chupin, Cucuzzella and Helal (2015).
8 See also Larson (1994): 'built or unbuilt, the projects ranked in an important competition are published, diffused, examined, discussed, and entered as credits in their authors' résumés.'
9 Cindy Walters, Director, Walters & Cohen, interviewed by the author, London, 14 October 2015.
10 See also Volker (2010; 2012).
11 See www.liverpoolphil.com and www.everymanplayhouse.com.
12 Sudjic (2006) observed that 'There is a perception that architect-dominated juries will favor a certain kind of design in which architectural expression is given a privileged position above all else […] There is also a suggestion that competitions should seek to select architects, rather than specific designs', and, according to Bergdoll (1989) 'Already in the first recorded competition [in 448 BC for a war memorial on the Acropolis], the idea of a competition as a democratic procedure for selecting a design confronted the opposing

reality of the specific competencies needed to judge architectural representations.' The issue, however, in the context of the two projects discussed in this chapter, is whether or not the involved parties selecting architects as part of an OJEU process, have or had an interest (and thus a responsibility) beyond commissioning the project, and if this has had an impact on the award.

13 Steve Tompkins, Director Haworth Tompkins Architects, interviewed by the author, London, 30 April 2015.
14 Haworth Tompkins had previously done projects for the Royal Court Theatre, the Almeida, the Theatre Royal Bath and the North Wall Performing Arts Centre, St Edward's School, Oxford, but the Everyman was their first entirely new build theatre. In addition the practice brought, at the time of tender, experience from other sectors such as residential, art galleries, office and administration buildings. Caruso St John were experienced in working with projects funded by the arts council, such as Tate Britain and the V&A Museum of Childhood in which they were also working with substantial amounts of existing and listed building stock. They had delivered one project that 'functionally' resembled the Liverpool Philharmonic Hall brief in their portfolio: the restructuring of the Barbican Concert Hall (2000/2001). The practice's portfolio at the time of tender also included residential and office buildings. In Haworth Tompkins' case it could be argued that it was their record of delivering technically very demanding theatre projects that made them an attractive choice for the shortlist, whereas Caruso St John's expertise with the refurbishment and remodelling of existing buildings, technically equally difficult and demanding, convinced their client to invite them for interview.
15 He adds the caveat, however, regarding open anonymous competitions that 'We would never do one now. It's not really worth our while doing competitions for very ordinary buildings with very ordinary juries, we're not going to win those competitions. We're trying to do special projects, we're not a huge practice, yet we have a very peculiar structure that's about the project we want to do, so we're much better off doing competitions for special buildings, that have people in the juries who we respect and who we know like our work.'
16 All of the shortlisted practices had experience in at least two other sectors, meaning that while they had specialist skills and knowledge of performance venues, they were also able to demonstrate that they could work in a diversity of sectors. While most architects will argue that specialist knowledge is good but by no means essential, particularly first time clients often find it difficult to award a project to a practice that has not got 'relevant' experience.
17 From the Everyman shortlist Haworth Tompkins and Ian Ritchie both had a broad portfolio covering a variety of sectors. Keith Williams and LDN Architects also brought experience from various sectors. Arts Team, now part of Aeadas, have completed a significant number of theatre and concert venue projects, and most of their portfolio consists of buildings for the performing arts.
18 Emma King, Capital Development Director, Liverpool Philharmonic, interviewed by the author, Liverpool, 8 July 2015.
19 For the Everyman architects and consultants were interviewed separately and the whole design team was subsequently assembled by the client. For the Liverpool Philharmonic Hall Caruso St John were appointed as lead consultants with subsequent responsibility for their chosen design team which was interviewed and appointed at the same time.
20 The idea that new talent is being discovered (Sudjic, 2006) through the system of design competitions is therefore slightly undermined in scenarios where selected practices are actively invited to participate in a specific competition.
21 In the above mentioned example, what King witnessed was a client organization, who did not have any expertise to judge a competition, and was 'led by the nose by people who only had a very short term relationship' with the client. A jury, composed of experts from the RIBA, visiting architects and members of the client's board, was swayed by the visiting experts who 'took the organization with them'.
22 Together with King the panel also included the Liverpool Philharmonic's chief executive, Michael Eakin; Simon Glinn (*at the time* Executive Director Liverpool Philharmonic Hall and Events – nb *this post no longer exists*); Tim Johnston (*at the time* Deputy Chair Royal Liverpool Philharmonic Board and member of the organization's Capital Development Board); and Christopher Allen (*at the time* Head of Fundraising).

23 King: 'They had a tour around the building, and that was our opportunity to just suss them out and to see what they were like just as we were walking around with them. We did the interviews on the stage, partly to scare them, because I think that's an important part of the interview process, and partly so members of staff could sit in and listen.'
24 The Everyman panel was also composed of members with a long term interest in the project: Prof. Michael Brown (Chair LMTT & Liverpool John Moores University); Rod Holmes (LMTT Board Member & Grosvenor); Deborah Aydon (Executive Director and Company Secretary); Gemma Bodinetz (Artistic Director); Robert Longthorne; Michael Hicks – Co-opted Member of Steering Group (Bruntwood Limited); John Clarke (Bucknall Austin). Source: Liverpool & Merseyside Theatres Trust. New Liverpool Theatres. Tender Instructions.
25 Emma King: 'It's a commercial decision practices make on whether they want the job or not. It's a fine balance for a client organization like this, saying fees aren't actually that important, it's the architect that's important. Actually we have to pay attention to the fee, we can't just ignore it. The fees were certainly important, but they weren't the most important in terms of the percentage scores. And Caruso St John weren't the cheapest and we did the tender as "lead consultant". So the architect has a number of other disciplines to work for them.'
26 Raymond Gilroy, Construction Director, Gilbert Ash, interviewed by the author, Liverpool, 21 September 2015.
27 Robert Longthorne, Building Development Director, Everyman and Playhouse Liverpool, interviewed by the author, Liverpool, 9 April 2015.
28 Garrett (1981) distinguishes between three different methods of consultation between client, problem owner (end user) and consultant (architect): expertise, process and contingent consulting. Barrett and Stanley (1999) explain (after Garrett), that expertise consulting 'stresses the relationship between the consultant and the client, excluding [...] the end user', that process consultancy is based on the relationship between consultant and problem owner (end user), and that in contingency consulting the consultant arbitrates between client and problem owner (end user) 'on the assumption that most of the information needed for an optimum solution already exists within the organisation'. Furthermore, 'only users appreciate the fine details of their work and so they are best placed to ask for changes', and 'if users are asked [...] they will feel part of the decision-making process and will "own" the project'.
29 Interestingly, prior to the Liverpool Philharmonic project being launched, the existing bar served as something of a precedent for the Everyman, and subsequently the new theatre building ended up with it's own version of a first floor bar facing Hope Street.
30 Longthorne calls the people represented on the screens and with whom the Everyman remains in touch, 'great ambassadors, those 105. Actually it is 106 now, one died, we had one birth. Someone gave birth after they had their photo taken and then there was someone else who didn't know they were pregnant when their photo was taken and they've given birth themselves now. So the whole thing has got a life of its own now, which is bizarre.'

References

Barrett, P., and Stanley C. (1999). *Better Construction Briefing*. Oxford: Blackwell Science.
Bergdoll, B. (1989). Competing in the Academy and the Marketplace: European Architecture Competitions, 1401–1927. In H. Lipstadt (Ed.), *The Experimental Tradition* (pp. 21–51). Princeton: Architectural Press.
Chupin, J.-P., Cucuzzella, C., and Helal, B. (2015). *Architecture Competitions and the Production of Culture, Quality and Knowledge*. Montreal: Potential Architecture Books.
Garrett, R. (1981). Facing up to change. *Architects Journal*, 28 October, pp. 838–424.
Larson, M. S. (1994). Architectural competitions as discursive events. *Theory and Society*, 23(4), pp. 469–504.
Lipstadt, H. (1989). The Experimental Tradition. In H. Lipstadt (Ed.), *The Experimental Tradition* (pp. 9–19). Princeton: Architectural Press.
Lipstadt, H. (2006). The competition in the region's past, the region in the competition's future. In C. Malmberg (Ed.), *The Politics of Design: Competitions for Public Projects* (pp. 7–27). Princeton: Woodrow Wilson School of Public and International Affairs.

Mirza & Nacey Research (2012). Procurement Case Studies, RIBA, London, 2012.
Mirza & Nacey Research (2012). Procurement Survey, RIBA, London, 2012.
Schmiedeknecht, T. (2005). Germany – (Un)edited architecture; Wettbewerbe Aktuell. In P. Davies and T. Schmiedeknecht (Eds), *An Architect's Guide to Fame* (pp. 121–142). Oxford: Elsevier / Architectural Press.
Schmiedeknecht, T. (2007). Karle / Buxbaum: the ordinary in procurement and design. *Architectural Research Quarterly*, 11(1), pp. 16–35.
Schmiedeknecht, T. (2011). Routine and Exceptional Competition Practice. In M. Roenn, R. Kazemian and J. Andersson (Eds), *The Architectural Competition – Research Enquiries and Experiences* (pp. 153–177). Stockholm: Axl Books.
Schmiedeknecht, T. (2013). Conventions of a competition system. *Architectural Research Quarterly*, 17(2), pp. 177–187.
Short, C.A., Barrett, P., and Fair, A. with Sutrisna, M., and Artopoulos, G. (2011). *Geometry and Atmosphere – Theatre Buildings from Vision to Reality*. Farnham: Ashgate.
Strong, J. (1976). *Participating in Architectural Competitions*. London: The Architectural Press.
Strong, J. (1996). *Winning by Design*. Oxford: Butterworth Architecture.
Sudjic, D. (2006). Competitions: The Pitfalls and the Potential. In C. Malmberg (Ed.), *The Politics of Design: Competitions for Public Projects* (pp. 53–66). Princeton: Woodrow Wilson School of Public and International Affairs.
Volker, L (2010). *Deciding about Design Quality*. Leiden: Sidestone Press.
Volker, L (2012). Procuring architectural services: sense making in a legal context. *Construction Management and Economics*, 30, pp. 749–759.
Ware, W. R. (1899). Competitions. *American Architect and Building News*, 66, pp. 107–112.

11 Advanced structural engineering

An interview with Werner Sobek

The third section of the book at hand has primarily discussed how architectural competitions shape subsequent phases of the building process. The two case studies presented have revealed that the competition does not necessarily produce stable results. There are, in fact, cases where unexpected events or changing circumstances entail significant shifts in the building project thereby challenging the findings gained in the competition. To further pursue this issue, we interviewed Werner Sobek, an architect and structural engineer (a profession that normally becomes involved in the building process only after a competition has been decided). From this perspective, he is an interview partner who reflects on the problem of entering the building process at a rather late stage. In the interview that closes part three of this volume, Werner Sobek discusses the value a structural engineer's early intervention can add to a project as well as the potential of being part of a competition team.

An interview with Werner Sobek, founder of the WERNER SOBEK GROUP

The WERNER SOBEK GROUP has realized a large number of projects via active participation in architectural competitions. Examples include: the Mercedes-Benz Museum in Stuttgart that opened in 2006 and was designed by Dutch architects UN studio; also in 2006, the Suvarnabhumi Airport Bangkok with the Chicago office of Helmut Jahn; Vakif Bank in Istanbul with Tabanlioglu Architects to be completed in 2017; as well as the National Museum of Qatar with Atelier Jean Nouvel, which is currently under construction and scheduled to open in 2017.

Between 1974 and 1980, Werner Sobek studied structural engineering and architecture at the University of Stuttgart, Germany. Subsequently, he worked as a post-graduate fellow on the research project 'Wide-Span Lightweight Structures', culminating with a PhD in structural engineering in 1987. Sobek founded his company in 1992. Today, the WERNER SOBEK GROUP has around 280 employees with offices in Stuttgart, Dubai, Frankfurt, Istanbul, London, Moscow and New York. He was Mies van der Rohe Professor at the Illinois Institute of Technology in Chicago from 2008 to 2014. Since 1995, he has headed the Institute of Lightweight Structures and Conceptual Design (ILEK) at the University of Stuttgart as a successor to architect Frei Otto and structural engineer Jörg Schlaich. He has founded several non-profit organizations such as the German Sustainable Building Council (DGNB), the Association for Architecture, Engineering, and Design (aed) and the Stuttgart Institute of Sustainability (SIS).

An interview with Werner Sobek

Werner Sobek (WS) was interviewed by Ignaz Strebel (IS) and Jan Silberberger (JS).

IS: Tell us about the competitions you are currently working on.

WS: There is a huge variety of competitions we are currently working on. One is a private competition for a detached house in Volgograd. The client invited four architects from all over the world. After this we have a competition together with Christoph Ingenhoven that concerns an extension to the Cologne Trade Fair. Together with Zaha Hadid we are taking part in two competitions in China; we are doing another competition with Helmut Jahn in America. With Ben Van Berkel we are working on a competition in Qatar. We are currently working on about twenty competitions, but it is just this project in Volgograd where we are doing architectural design. In the others we are responsible for the engineering of structures, façades and energy systems. In these cases, our input is only about 20 to 40 per cent in relation to the workload of the architects involved.

JS: Can you estimate how many of your projects derive from competitions?

WS: That is a large amount, certainly 70 per cent.

IS: Have competitions always played an important role in your work or is that something that has developed only over time?

WS: In the beginning competitions played an even more important role. When we say competition, we mean all forms of competition. This can also be an invitation from a client who approaches two different architects and casually asks them to produce sketches. Competitions were always important to us. However, the type of procedure that is of interest to us has shifted from anonymous, open competitions to invitation procedures or even personal calls.

JS: Often the structural engineer enters rather late into the process, that is, when the architect has already decided on decisive aspects of the design. How do you organize cooperation with architects?

WS: In general, the structural engineer enters the process after the draft is essentially determined, that means usually after the competition is decided. Sometimes it is explicitly stated in the competition programme that architects have to team up with structural engineers; but this is very rarely the case. We, however, insist that we participate from the very beginning in the design process – a stance that is highly appreciated by our business partners. They know that with us, they get exactly what they need to win the competition. Since we understand the language of architects we are able to give the appropriate answers. This allows for collaborations that incorporate the expertise of different disciplines in a way that reduces complexity instead of increasing it, which in turn accelerates the process instead of slowing it down.

JS: It is often argued that competition entries are sugar-coated in the sense that, for example, details with regard to the loadbearing structure, which impair the appearance of the design, are simply neglected. Then, after the competition, the structural engineer comes into play faced with the task of saving such drafts until it comes to planning permission. Are you familiar with such cases from personal experience?

WS: If you are preparing a competition entry it is only human that there are aspects of the design that you are not completely sure about. It is also quite human that you try to shift the focus away from such lacks of clarity. To give an example: we are currently working on a large airport terminal. We had not been involved in the competition but came in afterwards. So there is this hall with spans of approximately 200 × 200 metres without supports. When I saw that I said that is simply impossible. So we called the architect and he told us: 'Yes, there are concrete columns, at a distance of 12m.' Then we took another look at the visualization. Knowing now that there had to be columns, we could actually recognize them. They were rendered in such a transparent way that they were hardly visible. So we made them visible – and are now trying to develop a way of realizing the structure with far fewer columns than originally intended by the architect. Behaving as the architect did is certainly human. But at the same time it's a problem. You can only solve such a problem if you put together a brilliant team. Take the stadium that was built for the Olympic Games in Munich in 1972. (All judges thought that the winning design was by Frei Otto, yet it turned out Behnisch had designed it.) Anyway, at first all experts said that it was impossible to build the project the way it was shown on the plans. There were many counterproposals. Some wanted to build it using pre-stressed concrete, others wanted to build in wood. However, the client wanted the building realized exactly in the way it was shown on the plans, showing a new image of Germany – lightweight, transparent, with new colours and new forms. So they compiled the best team in the world: Fritz Leonhardt, Jörg Schlaich, Frei Otto, Günter Behnisch, Fritz Auer and Carlo Weber – and they got the project built. It worked. That means: if the idea or the proposal is not entirely abstruse, the power of a brilliant team, their knowledge and intellect, make such projects possible without significantly exceeding the calculated construction costs. If you want to push the boundaries you need to combine experts from the different fields of the building trade. Only in this way are you able to extend horizons.

IS: How do you organize the collaboration when you are working together with an architectural consultancy on a joint project?

WS: What is important – and this is our strength that distinguishes us from most other structural engineering firms – is that we understand the language of architects. In fact, I did not only study architecture, I also frequently dealt with fashion, aircraft and car design. The fact that we collaborate from the very first minute means that we influence the architectural design to a rather large extent. And vice versa: the architects significantly contribute to the design of the loadbearing structure. But such interaction – of mutually criticizing and challenging each other, of iteratively developing a design – is only possible if you can be sure that you as a structural engineer are also awarded a contract in case of winning the competition. That is, you have to have relatively good prospects for success and you have to be sure that the architect you are collaborating with is strong enough to ensure that you will be part of the further building process. You simply cannot consult an architect in an open competition with 200 participants investing hundreds of hours. This means that the designs in such open competitions are usually developed without in-depth expertise regarding structural engineering.

Now you can say that an architect is a generalist who is able to cover that. That may be true to a certain extent, but an architect can never be avant-garde when it comes to structural engineering. If you want to push the limits, you need to be a brilliant architect, but you also need avant-garde partners, an avant-garde structural engineer and an avant-garde energy expert. Without such cooperation you will always remain within the realm of what is already known, of what has already been done.

IS: We assume that young, inexperienced architectural offices also ask you to collaborate with them. How do you proceed in such cases?

WS: Competitions should indeed be an opportunity for young offices, but often they drive these young offices to the brink of ruin. They simply overreach themselves. They take part in a competition and win due to the charming and convincing entry they produce – but then the moment of truth comes. You simply are not able to handle a big project as an office consisting of only two talented young architects recruiting ten students. All the stuff that you do not necessarily need to know to win a competition comes crashing down around you during the project phase. Young offices then compensate for this lack of knowledge with additional work, commitment and frustration – but only partially. It is exactly at this point that we can support them and provide them with additional strength in terms of experience and size.

JS: You have founded several student competitions. What role do competitions play for students?

WS: More than fifteen years ago I founded the 'Stuttgart Lightweight Structures Award' to promote and reward innovations in the realm of lightweight construction – when speaking of 'lightweight' we do not only mean few resources but also little grey energy, low energy consumption in the utilization phase and 100 per cent recyclability. Another student competition that I founded is the 'Textile Structures for New Building', which concerns the application of technical textiles in construction. And the third award that I have founded is related to the initiative 'Architecture, Engineering and Design' (aed), which organizes about twenty events each year, typically all sold out. Within this initiative we organize a competition called 'neuland' [new territory], which explicitly asks for interdisciplinary contributions. With neuland we aim to address societally important issues, burning questions.

IS: Such as, for instance, the much-discussed question of energy?

WS: Indeed. You see, usually buildings are considered to be self-contained systems. But that is a mistake. We could also say that two buildings constitute one system. This is what I have done recently, in our project Aktivhaus B10 in Stuttgart's well-known Weissenhof Estate. Here a new construction together with an old building forms one composite system. That is, one building is strong with regard to energy efficiency – it produces more energy than it needs – while the other is weak; taken together, the two balance each other out. The overall societal demand calls for us to use as few resources as possible, and to produce as few CO_2 emissions and other waste as possible. That can be met by coupling such systems.

IS: To get back to architectural competitions: what is important is defined in the brief. Do you agree?

> WS: That's obviously true. However, this is a different level and a different problem. I would put it this way: the thicker the brief, the less it is read. And I do not believe that it is necessary to check how avalanche protection is ensured or where the doorknobs are placed when choosing a competent architectural firm. But such questions enter competition briefs because of a professionalization of the competition, that is, due to the fact that more and more clients assign a third party to organize their competition. It is often this third party which, in its meticulous attention to detail, forgets to create scope for the competing architectural consultancies. For the competition as such, too many details are, in my opinion, rather obstructive. They often prevent high-quality architecture.

JS: Is there, in general, a correlation between architectural competitions and the quality of the built environment?

> WS: I wouldn't say so. Take the development of Potsdamer Platz, for example. I observed it myself while we were still a very young office. For each construction plot a separate competition was organized and the executing architectural teams did not talk to each other afterwards since this was not intended by the City of Berlin. They did not want the architectural teams to communicate, and possibly build a collective strength against the city's senator for buildings. So they kept them all apart in order to control them better. But the result is a brick building from Kollhoff standing beside a glass tower designed by Helmut Jahn, next to a terracotta-turret by Renzo Piano. An urban ensemble has been generated that is of rather minor quality. One can only wonder how a city can blow such a chance. But this is exactly what happens if you run a competition for sub-area A and one for sub-area B and another one for sub-area C, and nobody knows what the others are doing. At Potsdamer Platz there are some very significant, fundamental problems. For example, paths in one complex do not coincide with paths in the adjacent complex. You leave one complex – all very good, everything nicely done there – you cross the street and then there is this offset that prevents you from simply entering the next building. The most trivial requirements were not met with regard to urban planning but also with regard to communication.

Index

Page numbers in *italics* refer to figures, those in **bold** refer to tables.

Actor-Network Theory (ANT) 20
Adams, John 118–9, 129
architectural competitions: and the discourse of architecture 152–3; plan–action relationship 21; process of 1; scholarship on 31–2
assemblage (building process) 20–1

Balogun et al. (2008) 62
Basel Kunstmuseum, Burghof Extension competition: brief 90; diagram providing essential information on the connecting wing *92*; evaluation and assessment criteria 87, *89*; the New Building *100*; overview of 87, 89–90; project perimeter *91*; 'Projekt 2: Neunhundertdreiundvierzig' (*Entry#2*) 91–5, *93*, 97–100; spatial specifications 88
Baumschlager Eberle 131; *see also* Eberle, Dietmar
Bergdoll, B. 38
BIG (Bjarke Ingels Group) 52, 53–4, *55*
Boltanski, L. and Thévenot, L. 66, 72
building processes: as assemblage 20–1; as chain of translation 20; and the competition process 5–6, 14, 19–20, 139, 149–50; construction projects, study 140–4, **141–2**, *143*; phases, subphases and goals of **16**, 139–40, *140*

Callon, M. 146, 148
calls for tender: the brief 69; and honourable mentions 79–80; and local knowledge 34; performance-orientated tender 35; research study 63–6, **65**; restricted tender procedures 64, 120; solution-orientated tenders 35; *see also* Official Journal of the European Union (OJEU) procedures
Carlsberg City redevelopment: challenges to the brief 110–11; competition format 103–4, 107–8, *108*; jury board deliberations 111–15; master plan for 108–9, *109*; methodology and data collection (research project) 109–10; use of dialogue in the competition 103, 104, 106, 107
Caruso, Adam 155, 156, 159, 164
chains of translations 20; *see also* moments of translation
Chupin, J.-P. 13–14, 15, 32
Chupin, J.-P. and Cucuzzella, C. 14
client requirements: absence of during design process 12; and anonymity of submitted proposal 4; client-architect relationship 132–3; and competition briefs 69–70; and the decision-making process 75; and open competitions 4–5; relationship with the design team, OJEU procedure 159–64
Cohen et al. 105
competition briefs: ambiguous 46; client requirements in 69–70; and the design process 13, 82, 83; honourable mentions 3; and the judging process 21–2; level of detail in 5, 46; Malcolm Reading on 81–2; Maritime Museum of Denmark 49–50; Mumbai City Museum North Wing 82; overloaded competition specifications 133; overview of 87; for solution-orientated competitions 87; ways of reading 13; Werner Sobek on 175
competition studies: before and after studies 17; archival approaches 12, 14–15; competition documentation 6–7; competitions in the making 12–15; critical approaches 7; ethnographic approaches 12–13; general criticism of competitions 8; good competitions 7–9; quality through competition 16–19; social production of architecture 9–11
competition tradition 72–3

decision-making processes: assessment criteria 74; client requirements 75; and dialogue-based criteria 103, 104,

106, 107; garbage can decision making process 105; justification processes 72–3, 74; and legitimisation of final choice 105; procedural legitimacy 56–7; and procurement law 69, 70–1, 72, 73; reading the decision task 67–70; research design and methods (public authority procurement) 63–6; role of expertise 74; sense-making process 60–1, 62–3, 66–7, **68**; and the spatial arrangements of jury venues 36–9, *37*; speculation and risk 118–20; substantive legitimacy 55–7; writing the decision process 70–2; *see also* jury boards

design process: and a building's contribution to public space 133–4; and the competition brief 13, 82, 83; and the competition process 6, 16–17, 19, 158–9; design component, OJEU tenders 152, 153, 154, 160–1, 165–6; jury boards as co-designers 14, 32; operability and usability 17–19; translation of future building in 45–6, 54–5; value of 153

digital tools 40–1

Eberle, Dietmar: career 131; on client–architect relationships 132–3; on competition programmes 133–4; on participation in architectural competitions 132; on the role of the jury board 132, 134–5

European Union (EU): Directive 2014/24/EU (public procurement) 2, 3; procurement regulations 59, 79; *see also* Official Journal of the European Union (OJEU) procedures

evaluation criteria: Basel Kunstmuseum, Burghof Extension competition *89*; and competition criteria 105; and jury judgement 105; ongoing development of 104–5

Everyman Theatre, Liverpool 155, 159, 161, *162*, 163–4, 165

exceptional competitions 34–5

The Experimental Tradition exhibition, National Academy of Design 9

fees: and EU Directive 2014/24/EU 2, 3; and OJEU procedures 160; and public procurement 2–3

France: architectural competitions in, historical 10, 38; Salle Melpomène, Ecole des Beaux-Arts 38, 39, *39*

Frey, P. and Kolecek, I. 12

geographical logics: the field of participants 33–6; spatial relations of the jury adjudication space 36–41

Gioia, D. A. and Chittipeddi, K. 62–3

honourable mentions 3, 79–80

interview stages: dialogue stage, Carlsberg City redevelopment 103, 104, 106, 107; Official Journal of the European Union (OJEU) procedures 153, 159–60

invitation competitions 1, 120, 157

jury boards: as co-designers of the winning project 14, 32; and the construction of commensurability 54, 57; decision-making process 13–14, 21–2, 32, 38–9, 61–2; Dietmar Eberle on 132, 134–5; and the digitalization of competition procedure 40–1; as domain specific experts 62; ethnographic research on 13–14; evaluation vs. judgement 135; jury rooms as places of knowledge 40; jury work as collective sensemaking 97–100; legitimisation of illegitimate process 54–5, 57–8; for open competitions 4; predictions for the future performance of the project 117–18, 119–20; pre-selection process 36–8; process of as unreplicable 40; solution- vs. performance-orientated solutions 35; spatial arrangements of jury venues 36–9, *37*; *see also* Basel Kunstmuseum, Burghof Extension competition; decision-making processes

jury reports 4, 22, 38

Kalkbreite construction project: competition for **141**, 144, *145*; main entities 146; moments of translation applied to 146–8, **147**

Kazemian, R. and Rönn, M. 70

Kenyon, Dan 164

King, Emma 158, 164, 165

KONKURADO – Web of Design Competitions 15

Kreiner, Kristian 13, 105

Larson, 153, 165

Latour, Bruno 12, 20

Lave, Jean 106

Lemaistre, Alexis 38, *39*

Lin, Maya 11

Lipstadt, Helene 9–10, 152, 154, 160

Liverpool Philharmonic Hall 155, 156, 157, 159, 160, 161, *162*, *163*, 164, 166

local knowledge: and calls for tender 34; and the field of participants 35–6, 42; and routine competitions 34

Longthorne, Robert 160, 161, 163

Maitlis, S. 63

Malcolm Reading Associates (MRA): 'after-care' provisions 83; global scale of 79;

Index

overview of 78–9; post-competition advisory roles 82–3; *see also* Reading, Malcolm

Maritime Museum of Denmark: budget and design considerations 52–3, 55, 57; competition controversies 50–1; competition design 49; competition overview 48–9; legal processes 51–2; study methodology 47; substantive legitimacy 56; valuation of creativity and architectural legitimacy 53–4

Meier, Richard 117, 129

Menz, S. 139

moments of translation: applied to the Kalkbreite construction project 146–8, 147; applied to wider study projects 148–9; chains of translations 20; overview of 146

Muhm, A. 139, 143

Mumbai City Museum North Wing, India 82

Nasar, Jack 17–19, 20

National Academy of Design, New York 9

Official Journal of the European Union (OJEU) procedures: costs and fees 160; design component 152, 153, 154, 160–1, 165–6; interview stage 153, 159–60; as process driven 165; public access to 154; tender process 157–9; track record and past client reception 151, 154; in the United Kingdom 151; and young practices 155

open competitions: anonymity of submitted proposal 4, 106, 157; as architecture as product 165; client requirements 4–5; custom-made procedures 5; Dietmar Eberle on 132; jury boards 4; as problematic 151–2; procedures 1, 120; and young practices 155

Opéra de la Bastille (Paris Opera House) 117, 129

Ott, Carlos 117

places of knowledge: jury rooms as 40; scientific laboratories 39–40

Potsdamer Platz, Berlin 175

process competitions 106

procurement regulations: and decision-making process 69, 70–1, 72, 73; EU procurement law 59, 79; Government Procurement (GPA) (WTO) 32, 33, 34; United States of America (USA) 79; *see also* Official Journal of the European Union (OJEU) procedures

public procurement: and architectural competition 3–4, 60; Directive 2014/24/EU 2, 3; EU procurement law 59, 79; and fees 2–3; global scale of 33; in Switzerland 3, 32–3; and World Trade Organization Agreement on Government Procurement (GPA) 32, 33, 34

Rafaeli, A. and Vilnai-Yavetz, I. 96

Reading, Malcolm: on competition briefs 81–2; first stages of the competition 81; on honourable mentions 79–80; relationships with the jurors 80–1; on young vs. established firms 80; *see also* Malcolm Reading Associates (MRA)

restricted competition 64, 120

risk 118–20

risk speculation: and decision-making process 118; financial preoccupations 126, 127; intellectual risk 127–8; nature of the speculation 129; vignette comparison 121–7

routine competitions 34

Sagalyn, L. B. 11

Schmiedeknecht, T. 34

Schön, D. A. 96

sense-making processes: in the decision-making process 60–1, 62–3, 66–7, 68; jury deliberations, Basel Kunstmuseum, Burghof Extension competition 91–5, 97–100; material dimension of 96; reflection-in-action 96–7; in slow-paced environments 96; Weickian theories of 95–6

Shapin, S. 39

situated learning 106

Sobek, Werner: on competition briefs 175; competitions and the built environment 175; on cooperation with architects 172–4; involvement in competitions 172; student competitions 174; and young practices 174

social mechanisms 45

'Solids IJburg' competition, the Netherlands 134

solution-orientated tenders 35; *see also* Basel Kunstmuseum, Burghof Extension competition

spatial specifications: Basel Kunstmuseum, Burghof Extension competition 88; Dietmar Eberle on 133–4

Spreiregen, Paul 7–8

starchitects 35

Stark, David 105

Stigliani, I. and Ravasi, D. 96

Stirling Prize 153
Strong, Judith 7–8
Suchman, Lucy 21
Sudjic, Deyan 117, 151, 157
sustainability 5
Switzerland: historical studies of competitions 12; KONKURADO – Web of Design Competitions 15; large architectural firms branches in 42; public procurement 3, 32–3; Regulation SIA 112, 139

tender process *see* calls for tender
Thompson, M. 127
Tompkins, Steve 155, 156, 159, 161, 163

United States of America (USA): competitions in 9; procurement regulations 79; Wexner Center for the Arts, Columbus, Ohio 17

Vietnam Veterans Memorial competition 9, 10–11

Weick, K. E. 95–6
WERNER SOBEK GROUP 171
Wettbewerbe aktuell (magazine) 17, *18*
Wexner Center for the Arts, Columbus, Ohio 17
World Trade Organization (WTO) 32, 33, 34

Zinn, Jens O. 119, 120, 129

Taylor & Francis eBooks

Helping you to choose the right eBooks for your Library

Add Routledge titles to your library's digital collection today. Taylor and Francis ebooks contains over 50,000 titles in the Humanities, Social Sciences, Behavioural Sciences, Built Environment and Law.

Choose from a range of subject packages or create your own!

Benefits for you
- Free MARC records
- COUNTER-compliant usage statistics
- Flexible purchase and pricing options
- All titles DRM-free.

Benefits for your user
- Off-site, anytime access via Athens or referring URL
- Print or copy pages or chapters
- Full content search
- Bookmark, highlight and annotate text
- Access to thousands of pages of quality research at the click of a button.

REQUEST YOUR **FREE** INSTITUTIONAL TRIAL TODAY

Free Trials Available
We offer free trials to qualifying academic, corporate and government customers.

eCollections – Choose from over 30 subject eCollections, including:

Archaeology	Language Learning
Architecture	Law
Asian Studies	Literature
Business & Management	Media & Communication
Classical Studies	Middle East Studies
Construction	Music
Creative & Media Arts	Philosophy
Criminology & Criminal Justice	Planning
Economics	Politics
Education	Psychology & Mental Health
Energy	Religion
Engineering	Security
English Language & Linguistics	Social Work
Environment & Sustainability	Sociology
Geography	Sport
Health Studies	Theatre & Performance
History	Tourism, Hospitality & Events

For more information, pricing enquiries or to order a free trial, please contact your local sales team:
www.tandfebooks.com/page/sales

Routledge
Taylor & Francis Group

The home of Routledge books

www.tandfebooks.com